W0227549

Talks About Wireless

*With Some Pioneering History
and Some Hints and Calculations
for Wireless Amateurs*

OLIVER LODGE

CAMBRIDGE
UNIVERSITY PRESS

CAMBRIDGE UNIVERSITY PRESS

Cambridge, New York, Melbourne, Madrid, Cape Town,
Singapore, São Paolo, Delhi, Mexico City

Published in the United States of America by Cambridge University Press, New York

www.cambridge.org
Information on this title: www.cambridge.org/9781108052696

© in this compilation Cambridge University Press 2012

This edition first published 1925
This digitally printed version 2012

ISBN 978-1-108-05269-6 Paperback

CAMBRIDGE LIBRARY COLLECTION

Books of enduring scholarly value

Technology

The focus of this series is engineering, broadly construed. It covers techno-logical innovation from a range of periods and cultures, but centres on the technological achievements of the industrial era in the West, particularly in the nineteenth century, as understood by their contemporaries. Infra-structure is one major focus, covering the building of railways and canals, bridges and tunnels, land drainage, the laying of submarine cables, and the construction of docks and lighthouses. Other key topics include developments in industrial and manufacturing fields such as mining technology, the production of iron and steel, the use of steam power, and chemical processes such as photography and textile dyes.

Talks About Wireless

In the 1860s, radio waves were predicted by James Clerk Maxwell in his work on electromagnetism. It took a further twenty years for the first experiments to produce a working demonstration. In this guide to radio technology, first published in 1925, eminent physicist Sir Oliver Lodge (1851–1940) provides a concise history of the development of the wireless radio, explains the theory behind it, and includes some practical tips for amateurs. Having lived through and contributed to the discovery, he explains the difficulty of the early experiments, which took place in a time when terms like 'frequency' and 'inductance', now taken for granted, did not exist in the scientific vocabulary. His first-hand account reveals the incredible efforts poured into the development of a revolutionary modern technology, rekindling the sense of wonder that once surrounded this strange new science.

Cambridge University Press has long been a pioneer in the reissuing of out-of-print titles from its own backlist, producing digital reprints of books that are still sought after by scholars and students but could not be reprinted economically using traditional technology. The Cambridge Library Collection extends this activity to a wider range of books which are still of importance to researchers and professionals, either for the source material they contain, or as landmarks in the history of their academic discipline.

Drawing from the world-renowned collections in the Cambridge University Library and other partner libraries, and guided by the advice of experts in each subject area, Cambridge University Press is using state-of-the-art scanning machines in its own Printing House to capture the content of each book selected for inclusion. The files are processed to give a consistently clear, crisp image, and the books finished to the high quality standard for which the Press is recognised around the world. The latest print-on-demand technology ensures that the books will remain available indefinitely, and that orders for single or multiple copies can quickly be supplied.

The Cambridge Library Collection brings back to life books of enduring scholarly value (including out-of-copyright works originally issued by other publishers) across a wide range of disciplines in the humanities and social sciences and in science and technology.

TALKS ABOUT WIRELESS

TALKS ABOUT WIRELESS

*With some Pioneering History and some
Hints and Calculations for
Wireless Amateurs*

By

SIR OLIVER LODGE

CASSELL AND COMPANY, LTD
London, New York, Toronto and Melbourne

First published 1925

Printed in Great Britain

PREFACE

THIS is a friendly rather than an ambitious book. It is a message of greeting to the great army of wireless amateurs and experimenters, from one who—always enthusiastic about ether waves—did some pioneering work; and who now admires the remarkable progress that has been made by others. May they all take it as conveying the author's good wishes, combined with a hope that, in the diversity of these gossipy chapters, each may find something acceptable, something worthy of his or her interest.

CONTENTS

INTRODUCTION

The Spirit of Progress

WHEN we think of the early beginnings of the generation and detection of ether waves (which, except in the form of light and other very high-frequency radiation, had never been generated or detected before 1887 and '88, though their theory was initially given by Clerk Maxwell in 1865), we may well be amazed at the rapid progress that has been made in their application to signalling all over the world, and especially to the transmission of human speech to surprising distances, with clear articulation and with avoidance of many of the difficulties which accompany cable transmission.

By no other means than through the free and untrammelled universal ether could the fine shades of articulate utterance be transmitted instantaneously, and with so little power considering the long range achieved in all directions.

The broadcasting of news and speech always seemed to be the object for which etheric waves were specially adapted ; but I never expected to live to see the process so successfully and marvellously developed, by the labour and ingenuity of a number of workers in co-operation, as has now been accomplished in a way which is a daily surprise and delight to listeners.

The creation of a new industry, which has come into existence as a consequence of the widespread demand, is but one of its benefits: the younger generation will surely be stimulated to take more interest than ever in the science of physics, which has rendered such an achievement possible. And the widespread and growing interest in science which has thus been begun may lead a few of the enthusiasts towards further investigation, and to further inventions and discoveries of which we at present have hardly an idea.

It is devoutly to be hoped that in the long run, when present international troubles have subsided, the power of rapid communication will surely conduce to better understanding among the nations, and will lead in due time to the much-desired but long-delayed era of universal peace. To this end much more than physical and material progress must contribute, for nothing can replace the whole-hearted desire for co-operative advancement of all nations on terms of mutual amity and good will.

The interchange of discoveries between the

nations has long been in operation. All scientific discoverers throughout the world virtually pool their resources and communicate to each other their results, except indeed those of a destructive and inhuman character. Secrecy is alien to the spirit of science ; and all true wealth increases in value when freely shared.

Indeed, that is the test of true wealth, as Ruskin long ago pointed out—namely, that the more it is shared the more it is possessed and the greater value it has for everybody. Witness the broadcast distribution of music and drama and works of art generally. These are not things for private possession only, but can be shared and enjoyed by all. Things of this sort constitute the true wealth of humanity, and to that category scientific discoveries belong.

The spirit of nationalism is wholesome enough if it be held reasonably and without flaunting claims to superiority. The spirit of emulation is also wholesome, for it is by no means the same as competition. Emulation is the desire to do something better than has been done before, and better than others have as yet learnt to do it ; while competition is mainly the effort to monopolize some activity, and to do things instead of others. In extreme cases unfair competition may prevent others from taking their due share in the beneficent work of the world.

Hence progress surely lies in the direction of co-operation all round, each nation and each individual doing his best, and not seeking to prevent others from doing theirs. Nations which interchange scientific discoveries might also wisely interchange commodities, since not every locality is equally suitable for producing everything. Division of labour, of this sane and salutary kind, might be recognized as good. And international jealousies, based on mere rivalry and competition, ought to cease, especially rivalry and competition in armaments and instruments of destruction.

Civilization ought to have progressed too far by now for the perversion of ingenuity responsible for the construction of diabolical and otherwise useless mechanism, and for the artificial increase of those natural evils among which it is our lot to live and against which it is our business to contend.

Sorrow and sighing there must be in plenty during this planetary existence, without our trying to increase their sum, without adding to each other's difficulties, and without bringing about those horrors of death and torture and destruction which, when they occur by inadvertence or by accident or by the unconscious and uncontrolled forces of Nature, arouse world-wide feelings of sympathy and desire to help.

It is not viciousness that perverts humanity most, it is a kind of stupidity, a lack of clear perception of what the straight course is; so it blunders into evil without really willing it, swayed by passing fancies, and lacking self-control.

> Our wills are ours, we know not how;
> Our wills are ours, to make them Thine,

and until the will of man is brought into harmony with what we can gather of Divine intention, humanity cannot be called really and effectively civilized, although it may achieve marvellous feats of locomotion, and although it has extended the range of human speech to distances undreamt of by our ancestors. Now that we are able to travel farther and faster, we should travel to some good purpose. And now that we can speak across a continent, let us see to it that we have something worthy to say.

P A R T   I

RADIO IN GENERAL

TALKS ABOUT WIRELESS

CHAPTER I

On Broadcasting

THE chief feature which distinguishes humanity from the rest of the animal creation is the power they have gradually acquired of communicating their experiences consciously and explicitly to each other and to subsequent generations. This is the basis of Education in its widest sense. Individuals are not limited to their own experience ; they have the benefit of the accumulated experience of their fellows and of all past generations since the dawn of civilization. Some kind of race memory seems to exist among animals too, but it is of a vague and indefinite and rather mysterious kind ; and if it can be called communication at all, it must be of an entirely unconscious order. Mankind, however, is able to transmit consciously information and experience. Contemporary information is conveyed through the newspaper press ; while historical and scientific information of the most varied

and multifarious kind is contained in books.
In other words, the art of speech and the per-
manent record of speech enabled by printing
render accessible more information than any
one individual can hope to assimilate. The
educated man is one to whom all these resources are open : but the least laborious method
of getting at it, to all except the serious student,
is through the ear rather than through the
eye ; in other words, most people prefer to
learn by hearing rather than by reading. The
living voice has a power of attracting attention
more vivid than the pages of a book.

Unfortunately humanity, having learnt to
speak, was not satisfied with a single code or
language ; but each large community developed
a language or code of its own : and accordingly
much misunderstanding exists between different communities, since they cannot freely
converse with each other ; and this must have
been, and still is, a fruitful cause of misunderstanding and disputes, and even of wars. This
planet is not a very large one ; and it may
reasonably be hoped that at no distant future
every part of humanity, every section of mankind, will have a better understanding of each
other's outlook and aims and ambitions and
hopes, through the medium of a common
tongue : and every step that is taken towards

intercommunion and the annihilation of distance must be of the utmost importance. Already books can be circulated all over the world, and by means of translation can be made accessible, even when their language is unknown. Moreover, the English-speaking race has spread over so great a part of the world that already the continents of America, Africa, and Australasia are inhabited in great part by English-speaking people, and thus communities and families spread over an immense area have friends and relations with whom intercourse is not only a matter of political and social understanding, but can be of a personal and domestic and affectionate character also.

The invention of the telegraph, and the laying down of cables, have, in the day and generation of many now living, come into active being, and have linked up the world in a way which was never experienced before. Not only information about the past is now accessible, but the living and daily experience of the present is also distributed to the ends of the earth.

Quite a short time ago the information so distributed was only accessible by reading; the transmission was effected by expert operators, and then found its way into print. Now, however, in the memory of those still young, a still more rapid and efficient mode of contemporary intercourse has become possible. The hearing of speech at one place is not limited to those within the reception of acoustic waves in the air ; but means has been found of converting those sound waves into ether waves, which, though they do not produce any effect on the human organism, and by our unaided senses are inapprehensible, yet by the use of instruments of reconversion the ether waves can be received and transmuted back into sound waves ; so that that wonderful instrument, the human ear, can receive them when thus reconverted. The human mind can thus interpret indirectly the ether waves, so as to extract from them information, in the same way as it used to interpret air waves in the case of the speech of neighbours and friends close at hand.

This is a step in advance of the utmost importance, and the consequences of it we do not yet fully foresee. This new method of communication is still in its infancy, and how it will develop no man can certainly tell ; but already we may reasonably look forward to the time when friends and relations in New Zealand will be able to communicate with those in the older countries, not through the medium of skilled operators and writing, but by direct speech. However that may be, we know already that the inhabitants of any one country can be linked together, irrespective of distance, in a way which was never before possible in the history of the world.

The extraordinary way in which the ether connects the worlds together, so that ether waves bring information about the physical and chemical constitution of distant stars, has been known for half a century, or to a minor degree for several centuries. The waves travel through the ether without loss, carrying their information with them, ready to be interpreted by any whose minds are open and who have the instruments necessary. The emission and reception of light is a very ancient system of broadcasting; the earliest receiving set is the human and animal eye.

INVENTION AND MEDITATION

The conscious utilization of ether waves is a modern and growing art, at first depending wholly on the human eye, enlarged and extended first by the telescope and then by the spectroscope. Twenty or thirty years ago no other instrument seemed possible : light seemed the only method of intercommunication through the ether. But the discoveries of Maxwell and

Hertz made possible the artificial production
of ether waves ; and the discoveries of some
still living made it possible to control and
modify those waves, so that they should have
any desired rate of vibration, and so that they
should be made to correspond with the intri-
cacies and peculiarities of the consonants and
vowel sounds which are used in ordinary speech.

Thus the present stage has been reached :

(1) By discovering how to produce long
waves ;

(2) By discovering how to receive them ;

(3) By discovering how to modify them ; and

(4) By discovering how to receive the modifications in detail.

In these stages the invention of the telephone, one of the most remarkable inventions
of last century, has played an extensive part.
But the real miracle is, as always, the human
ear and human eye, which are receiving instruments able to interpret mere vibrations or
tremors into actual thought. Not that the
instruments themselves affect the conversion :
there is a step here which is really mysterious
and which we do not understand, namely, the
way in which the physiological mechanism is able
to affect the mind. This is a mystery we have
grown so accustomed to that we fail to realize

the wonder of it. It is sometimes felt that wireless intercommunication itself is mysterious ; but it is no more mysterious than any other method of communication : in common speech we utilize the vibrations of the air ; in wireless speech we utilize the vibrations of the ether. The wonder is that we can utilize the vibrations at all, and that a thought can be conveyed from one mind to another by the motion and quivering of material particles.

This power, long possessed by the human race and utilized by them in a commonplace and customary manner, has now been extended. Some of the results we already begin to see. People living in remote country districts can now listen to the sound vibrations which are produced in any lecture or concert hall : and accordingly the entertainments of a town have become accessible to dwellers in remote valleys among the hills, or in the great plains of a continent. They think nothing about sound vibrations, nor about the ether waves into and from which those vibrations have been converted ; but they get the results. Hitherto life in the depths of the country has been felt by many to be isolated, dull, and monotonous ; it ought not to be so, and would not be so if people were educated, so that they could appreciate the marvellous

processes of Nature which are going on around
them ; they should be able to read and cul-
tivate their minds even to greater advantage
away from the distractions of a town ; but
for the majority of people that is too much
to expect, and accordingly there was a tendency to flock to the towns and desert the
country. There is some hope that that un-
economical and rather sad movement can now
be stemmed. Every village can now be supplied with sufficient light to make the winter
evenings tolerable, and can be provided with
sufficient entertainment through the labours of
those who are producing and broadcasting
speech and music. Poetry and literature can
be brought home to them in an easy way, and
even scientific information can be imparted ; so
that their minds need no longer lie fallow and
idle : they can have something to think about
which will relieve the monotony, and cause
them to take a rational interest in the Universe
and in the wonders of existence.

For existence itself is the greatest marvel : a
study of the processes by which it is continued
and adapted and beautified, and the attempt
to apprehend the meaning of it all—problems
which have long occupied the minds of philoso-
phers and students—should be, in their re-
spective and relative degree, a source of enjoy-

ment and recreation and rational meditation to every human being. It is easy to become supine, with mind closed to everything but material needs and comfort : it is easy, but it is unworthy, it is a reversion to the animal type. The animals must be mainly occupied with these things, but humanity should rise above them. Some outlook beyond seems possible even to the animals. What the lark means by his song we do not know ; it seems to be an instinctive outpouring of joy in existence—a joy which too many of mankind have lost, but which must sooner or later be recovered. Sooner or later a greater number of mankind will be satisfied with a moderate attention only to material needs ; they will feel that the cultivation of their souls is a worthier object ; they will learn that they have an immortal existence which cannot be really satisfied except by those faculties and emotions which from time to time, in the higher members of the race, have called forth spontaneous expression of worship, and which should arouse in all rational beings the feelings of wonder, love, and praise.

CHAPTER II

Early Pioneering Work in Ether Waves

NOW that electric waves are so easy to produce, and so absurdly easy to detect, it is amazing to think how long it took to discover them, and how many times they hinted at their existence before they were properly recognized and appreciated. From time to time in laboratories little residual phenomena appeared, which were noted as rather interesting and surprising but were not seriously followed up. They were sporadic phenomena, to which the observers had no clue, of which therefore they did not understand the meaning, and which accordingly were relegated for the most part to the subsidiary and unimportant rôle of accidental disturbances, a sort of *poltergeist* phenomenon, which could not be brought to book and which seemed rather lawless and capricious. Occasionally and for a short time some man of genius had a suspicion that the odd little sparks and things which he observed must have a meaning, which might be of importance, if it could be ferreted out ; but meanwhile the lack of a clue or any guiding theory caused them to be forgotten again and not followed up either by the observer or his contemporaries.

Early pioneering work is too often overlooked and forgotten in the rush of a brilliant new generation, and amid the interest of fresh and surprising developments. The early stages of any discovery have, however, an interest and fascination of their own, and teachers would do well to immerse themselves in the atmosphere of those earlier times, in order to realize more clearly the difficulties which had to be overcome, and by what steps the new knowledge had to be dovetailed in with the old. Moreover, for beginners, the nascent stages of a discovery are sometimes more easily assimilated than the finished product. Beginners need not, indeed, be led through all the controversies which naturally accompany the introduction of anything new ; but some familiarity with these controversies and discussions on the part of the teacher is desirable if he is to apprehend the students' probable difficulties. For though he does not himself feel them now, the human race did feel them at their first introduction ; and the individual is liable to recapitulate, or repeat quickly, the experience of the race.

A large number of people now interested in the most modern developments of wireless

have but little idea—perhaps none at all—of
the early work, in apparently diverse directions,
which preceded and made such developments
possible. No one, indeed, can subsequently feel
in touch with the history so closely as those
who have lived through the period covered
by it. Only those who have survived the
puzzled and preliminary stages of a discovery
can appreciate fully the contrast with subsequent enlightenment. It may suffice to say
that the term " inductance " or " self-induction," which we now use so glibly, did not at
first exist ; and that so late as 1888 Sir William
Preece still spoke of it as " a bug-a-boo " :
whereas it is the absolute essential to tuning,
and even to electric oscillation.

Lord Kelvin, who first introduced it in 1853 as
a mathematical coefficient, without any explanation, called it " electro-dynamic capacity."
The name self-induction was given to it by
Maxwell, though it was long before it was
understood or utilized ; and the name " inductance " is a nomenclature of Heaviside.

I wish here to say little about anything to
do with wireless later than 1896. What I have
to deal with is the early pioneering work apart
from practical developments. Let me say at
once, to avoid misunderstanding, that without the energy, ability, and enterprise of Signor

Marconi, what is now called "wireless" would not have been established commercially, would not have covered the earth with its radio stations, and would not have taken the hold it has upon the public imagination. Before 1896 the public knew nothing of its possibilities : and for some time after 1896, in spite of the eloquence of Sir William Preece and the demonstrations by Signor Marconi, the public thought it mysterious and almost incredible ; and still knew nothing about the early stages. Indeed, I scarcely suppose that Signor Marconi himself really knew very much about them. He had plenty to do with the present ; he felt that the future was in his hands ; and he could afford to overlook the past.

It may be doubted whether the younger generation, who are so enthusiastically utilizing and perhaps improving the latest inventions, will care much about the past either. Incidentally, however, I want to say two things to those who are occupied with the subject to-day First, do not hesitate to speak and think of the *ether of space* as the continuous reality which connects us all up, and which welds not only us but all the planets into a coherent system. Do not be misled by any misapprehensions of the Theory of Relativity into supposing that that theory dispenses with the ether merely because it succeeds in ignoring it. You can ignore a thing without putting it out of existence : and the leaders in that theory are well aware that for anything like a physical explanation of light or electricity or magnetism or cohesion or gravitation, the ether is indispensable. The ether has all these functions, and many more. We are utilizing it every day of our lives ; and it would be ungrateful as well as benighted if we failed to render due homage to its omnipresent reality and highly efficient properties. It lies at the origin of all electrical developments, and forms the basis for this new and broadcast method of communication.

That is one thing : the second is to congratulate all those whose wonderful and rapid advances have rendered possible the astonishing feat of, in any sense and by whatever means, carrying the human voice across the Atlantic. When Signor Marconi succeeded in sending the letter " s " by Morse signals from Cornwall or Ireland to Newfoundland, it constituted an epoch in human history, on its physical side, and was itself an astonishing and remarkable feat. The present achievement of changing over from Morse signals to ordinary speech, made possible by the valves of Prof. Fleming and Dr. Lee de Forest and others, is a natural though still surprising outcome and development of long-distance transmission, and must lead to further advances, of which at present we can probably form but a very imperfect conception.

Well, now I must go back to early times. In or about the year 1875 Mr. Edison observed something, which at that time could by no means be understood, about the possibility of drawing sparks from insulated objects in the neighbourhood of an electrical discharge. He did not pursue the matter, for the time was not ripe ; but he called it " Etheric Force "—a name which rather perhaps set our teeth on edge ; and we none of us thought it of much importance. Silvanus Thompson, however, took up the matter in a half-hearted sort of way, and gave a demonstration to the Physical Society of London in, I believe, June, 1876— a paper which I have had a little difficulty in finding in the Proceedings of that Society. Nothing much came of it, however, though his argument tended to show that the sparks could be accounted for on known principles. The value of this is merely that it must have rendered Thompson susceptible to methods of detecting real electric waves when they were discovered later.

Colonel Crompton informs me that these Edison sparks were also examined by Prof.

c

Elihu Thomson in the Central High School Laboratory, Philadelphia, in 1875 or 1876. Thomson used an ordinary Ruhmkorff coil, one terminal to earth—a water-pipe ; the other terminal being connected by a few feet of wire to a large tin vessel mounted on a glass jar. Whenever the coil sparked, he found that small sparks could be drawn from knobs and pipes in many other parts of the building, even up on the sixth floor, 100 feet away from his apparatus. He claimed that the reason Edison's etheric force did not affect galvanometers and electroscopes was because it was oscillatory ; the alternations of current being far more rapid than any instrument could respond to.

In fact, he was dealing with high frequencies before the word frequency was in electrical or optical use : the apparent lack of polarity exhibited in Edison's effect was accounted for. But Elihu Thomson did not follow up the observation : he did not recognize their Maxwellian significance. In fact, not till 1887 and 1888, when I independently stumbled on the same phenomenon, was it recognized that there —or rather in the extensions or expansions of that phenomenon—was something which could be detected and measured, running along wires with a definite wave-length, and that these impulses were truly Clerk Maxwell's waves ;

which were thus introduced to science on wires at about the same time as Hertz was detecting them in space. (See latter half of my paper in *Philosophical Magazine* for August, 1888.)

It was found afterwards that Joseph Henry, at the Smithsonian Institution in Washington, had observed something of the same kind so early as 1842. He seems to have had an intuition of the possible importance and far-reaching consequences of his observation, for he speaks as follows (I quote a passage cited in my lecture " On the Discharge of a Leyden Jar," printed at the end of my 1889 book, " Modern Views of Electricity ") :—

" It would appear that a single spark is sufficient to disturb perceptibly the electricity of space throughout at least a cube of 400 feet of capacity, and . . . it may be further inferred that the diffusion of motion in this case is almost comparable with that of a spark from flint and steel in the case of light."

That is to say, so early as 1842 Joseph Henry had the genius to surmise that there was some similarity between the etherial disturbance caused by the discharge of a conductor and the light emitted from an ordinary high temperature source.

In the light of our modern knowledge, and Clerk Maxwell's theory, we now know that the similarity is very near akin to identity. Both sources emit ether waves, though prodigiously differing in length.

Subsequent to these early stray observations an amazingly suggestive observation, of a partially similar kind, was made by that singular genius and brilliant experimenter David Hughes, the inventor of the microphone or telephonic transmitter, and of the Hughes Printing Telegraph still used in France. He was a man who " thought with his fingers," and worked with the simplest home-made apparatus —made of match-boxes and bits of wood and metal, stuck together with cobbler's wax and sealing-wax. Such a man, constantly working, is sure to come across phenomena inexplicable by orthodox science. As a matter of fact, Hughes unknowingly was very nearly on the trail of what was subsequently discovered, in a so much more enlightened manner, by Hertz. Hughes, too, got sparks in the course of his experiments, but he also got something very like coherer action by means of his microphone detectors. They enabled him to get actual galvanometer deflexions—such as Hertz never got.

All this was early in the 'eighties and before either Hertz or me. Hughes was a telegraphist, and though he would never have worked out the subject mathematically as Hertz did, and would not have been interested in matters of theory, he might well have stumbled, even at that early date, on something like a rudimentary system of wireless signalling had he been encouraged. But he was not encouraged. He showed his results to that great and splendid mathematical physicist Sir George Stokes; and Stokes, alas, turned them down, considering that they were explicable either by leakage or some other known kind of fact.

That is the danger of too great knowledge; it looks askance at anything lying beyond or beneath its extensive scope; whereas an experimenter operating at first hand on Nature may quite well occasionally stumble on a fact which lies outside the purview of contemporary science, and which accordingly neither he nor anyone else at the time understands. Crookes himself had a similar experience. In his pertinacious and systematic way he explored many unfamiliar and untrodden regions; and he also invited the attention of Stokes to a simple and easily investigated case of abnormal movement; who, however, perceiving that such motion was physically impossible, declined to take any interest in it or even to see it. His reason told him (and the reason he gave was) that on recognized principles the asserted phenomenon could not happen. But that was precisely its point of interest, and that was why Crookes with his instinctive sagacity conceived that such things held within them the germ of a great science of the future.

In Crookes's case the germ still remains unfructified by orthodox science. In Hughes's case the germ was rediscovered and has borne fruit a million-fold. But this is to anticipate. Suffice it now to direct attention to the collection of Hughes apparatus now unearthed by the energy and piety of Mr. Campbell Swinton, and exhibited in the Museum of Science at South Kensington. And let us try to avoid imitating the mistakes of our revered scientific ancestors : though I admit it is a difficult task. So much rubbish is brought to our notice that we are bound to run the risk of neglecting a jewel amid the chaff.

THEORIES OF LIGHT

These spasmodic observations, however, are not exactly discoveries : they were more akin to vague intuitions. The first and gigantic step in the real discovery was made by Clerk Maxwell, in or about 1865 : and he made it in mathematical form, not in experimental actuality, by one of those superhuman achievements which are only possible to our greatest mathematical physicists. He did not discover either the way to generate those ether waves, or to detect them ; but he did give their laws ; he legislated for them before they were born. He knew the velocity with which they must move, and gave implicitly, though without elaboration, the complete theory of their nature.

Up to his time the nature of light was unknown. All the other theories of light had attempted to explain it on mechanical principles, like the vibrations of an elastic solid. Light was known to consist of transverse waves : the wave-length and the frequency of oscillation could be determined. But no one knew what was oscillating, nor what the mechanism of propagation was. With extraordinary genius Fresnel and MacCullagh had explained the phenomena of light in all detail as regards reflection, refraction, diffraction, interference, and polarization. But the nature of the waves was unknown ; and the elastic solid theory, though fascinating, was felt by those who dived most deeply into it to contain some flaw, and to be, strictly speaking, unworkable. Light did not seem explicable on dynamical principles—the principles which were so fruitfully devised by Galileo and Newton for dealing with ordinary matter.

MacCullagh's theory indeed was not dynamical, and in that respect had some advantage. But it was also vaguer and less definite on that account; though, being thus indefinite and yet enabling results to be achieved, it was less liable to be upset and replaced by future discovery.

To Clerk Maxwell we owe the epoch-making discovery that light was not a mechanical oscillation at all, that the ordinary mechanical properties of matter did not apply to it, but that it was explicable solely and wholly in terms of electricity and magnetism. It is impossible to sum up his discovery in a few words; but roughly we may say that the most obvious outcome was:

(1) That if electric waves could ever be generated they would travel with the velocity of light.

(2) That light was essentially an electromagnetic and not a mechanical phenomenon.

(3) That the refractive index of a substance was intimately related to the dielectric coefficient.

(4) That conductors of electricity must be opaque to light.

Maxwell showed further, though he did not then express it in language of this character, that the ether had two great and characteristic constants, of value utterly unknown to this day, though guessed at by a few speculators like myself—one of them the electric constant of Faraday called " K " ; the other the magnetic constant of Kelvin called "μ." It was impossible then, and it is impossible now—though it is not likely always to remain impossible—to determine the value or even the nature of either of these constants. But Maxwell did perceive a way of measuring their product ; and he was the first to measure it. Their product is known ; and it is equal—as he showed it must be—to the reciprocal of the square of the velocity of light.

Well now, this great discovery aroused in us young physicists the greatest enthusiasm. In the early seventies of last century—I think about 1871 or 1872—I remember discussing it with the man we all now know and honour, J. A. Fleming, who at that time was a fellow-student with me in Prof. Frankland's advanced chemical laboratory at the brand-new College of Science, South Kensington. A year or two later, at Heidelberg, I studied Maxwell's treatise pretty thoroughly, and formed the desire to devote my life if possible to the production and detection of Maxwell's electric waves.

CHAPTER III

The Discovery of the Waves

I USED to discuss the possibility of producing these waves with my great friend, G. F. FitzGerald, whose acquaintance I made at the meeting of the British Association in Dublin in the year 1878 ; and he wrote some mathematical papers discussing the possibility of producing such waves experimentally. I myself also spoke at the British Association about them, in 1879, 1880, and again in 1882, at the Royal Dublin Society. FitzGerald, as I say, mathematically examined what then seemed the abstruse question of electric wave production ; and after some hesitation came to the conclusion that direct artificial generation of waves was really possible on Maxwell's theory, in spite of certain recondite difficulties which at first led him to doubt it. (See " Scientific Writings " of FitzGerald, edited by Larmor, pp. 90–101.) Indeed, one of his papers on the subject was originally entitled " On the Impossibility of Originating Wave Disturbances in the Ether by Means of Electric Forces." The prefix " im " was subsequently dropped ;

although his first, or 1897, paper concluded thus :

" However these [displacement currents] may be produced, by any system of fixed or movable conductors charged in any way, and discharging themselves amongst one another, they will never be so distributed as to originate wave-disturbances propagated through space outside the system."

In 1882 FitzGerald corrected this erroneous conclusion, and referred to some early attempts of mine at producing the waves. (" Scientific Writings," p. 100.) I state all this in order to emphasize the difficulty which in those early days surrounded the subject on its theoretical as well as on its practical side.

In 1883, at the Southport meeting of the British Association, FitzGerald took a further step and surmised that one mode of attaining the desired result would be by utilizing the oscillatory discharge of a Leyden jar, if only we had the means of detecting such waves when they were generated.

Inspired by FitzGerald's views, I and A. P. Chattock, now of Bristol, then working at Liverpool, did succeed in getting clear evidence of the existence of these waves, running along wires : and we were able to measure and express their length in metres or fractions of a kilometre, by the nodes and loops which accompanied the stationary waves caused by their reflection at the far end of the wire, and which displayed themselves to the eye in the dark by spaced-out brush-discharge luminosities, as well as in other more metrical ways. The arrangement was afterwards known by the name of Lecher, who made many measurements on this plan with the aid of vacuum tubes, as also did Dr. Dragoumis in my laboratory. So I summarize as follows :—

" In 1887 and 1888 I was working at the oscillatory discharge of Leyden jars (initially in connexion with the phenomena of lightning), and I then found that the waves could be not only produced but also detected, and the wavelength measured, by getting them to go along guiding wires adjusted so as to be of the right length for sympathetic resonance. Thus I obtained the phenomenon of electric nodes and loops, due to the production of stationary waves by reflection at the distant end, and in my own mind thus verified Maxwell's theory."

Transmission along wires popularly sounds different from transmission in free space, but it was well known to me that the process was the same, and that the waves travel at the same speed, being only guided by the wires, much as sound is guided in a speaking-tube, without the velocity of transmission being to any important extent altered. The theory is given near the end of my paper—an important one as I think, and as Silvanus Thompson agreed—in the *Philosophical Magazine* for August, 1888, where the experimental production of much shorter waves is also foreshadowed.

The beginning of my experiments was reported to the Society of Arts in April 1888 ; they are recorded in the *Phil. Mag.* for August 1888, and they were more completely described orally at the British Association at Bath that year. (See the *Electrician*, vol. 21, pp. 607–8, September, 1888.)

In that year, also, I heard for the first time of Hertz's brilliant series of experiments, where, by the use of an open-circuit oscillator, he had obtained waves in free space, and by reflection had also converted them into stationary waves and observed the phenomena of nodes and loops, and measured the wave-length.

Attention was directed to these experiments of Hertz by FitzGerald in his presidential address to Section A of the British Association meeting at Bath in 1888. No wonder they interested him ; for they showed that his method of utilizing the oscillatory discharge of a Leyden jar was effective ; and, to the surprise of all of us, including Hertz himself, that the waves

from an opened-out condenser (or early insulated aerial) had sufficient power to generate
sparks in an insulated conductor upon which
they impinged ; the detecting conductor, as
generally used by Hertz, being in the form of
a nearly closed circle with a minute spark gap
at which the scintilla appeared. The radiating
power of even a small Hertz oscillator was
calculated by me in a subsequent paper (*Phil.
Mag.* for July 1889, p. 54), and was found to
be 100 horse-power, while it lasted. The dura-
tion was excessively short, for, at that date,
practically all the energy was expended in a
single swing (about the 100-millionth of a
second), but its property of exciting little sparks
was amply explained by its remarkable radiating
power.

This work of Hertz was splendid. He was
then professor at Carlsruhe, still quite a young
man. He had been trained under Helmholtz ;
and I had made his personal acquaintance in
Berlin when I went to call on Helmholtz in
1881, on a tour of the universities of the
Continent. He was then Helmholtz's demon-
strator, and was thought highly of by that
great master. He could speak English, and
was very friendly. I did not see him again till
some time after the publication of his great
discovery.

Hertz was not at that time fully acquainted with Maxwell's theory, though indeed he knew his equations better than any other German except Helmholtz. Maxwell had not then made any serious impression on the Continent. Even Hertz does not seem at first fully to have realized what he was doing, and did not use the words " electric waves." That title was attached to his subsequently translated book at the suggestion of Lord Kelvin. He spoke about " the out-spreading of electric force " ; somewhat as Joseph Henry had done. That was the title of his book. He worked out the phenomena he observed with extraordinary skill, both experimentally and mathematically, rapidly perceiving that Maxwell's theory could be applied to it and that it might be elaborated in detail so as to include the whole of his phenomena. He it was who drew those accurate diagrams of the genesis of the waves, showing what is happening near the oscillator at every phase—diagrams which now appear in most textbooks and of which the upper half is represented as scouring across the country when an aerial is earthed. He knew that true waves were not emitted till beyond a quarter-wave length from the source. He knew how they were polarized, and how their intensity differed in the equatorial and polar directions, and how it varied with what may be called latitude. In fact, he rapidly came to know all about these waves.

As to us, we knew not which to admire most — his experimental skill when working with a tiresome and irritating mode of detection—for he had nothing but a scintilla seen in the dark, as a detector—or his mathematical thoroughness in ascertaining the laws of the waves' generation and propagation. A synopsis of his equations will be found clearly cited in Preston's " Theory of Light," as well as in other books. I translated some of his papers into *Nature*. Never was there the smallest iota of jealousy between us, or anything but cordial and frank appreciation. Maxwell and Hertz are the essential founders of the whole system of wireless. That is to say, they constructed the foundations solidly and well. Of the superstructure—splendid as it is now—we are as yet far from seeing the completion.

In March 1889 I lectured to the Royal Institution on " The Oscillatory Discharge of a Leyden Jar," and incidentally exhibited many of the effects of waves, both on wires and in free space, with overflow and recoil effects. But there was nothing akin to *signalling* exhibited in this lecture, as there was in the subsequent lecture in 1894.

Nevertheless, Sir William Crookes, on the strength of these experiments—which he mentions—wrote a brilliant article in the *Fortnightly Review* for February, 1892 (vol. 51, p. 173) in which he foreshadows actual telegraphic accomplishment by that means, and indicates also the possibility of tuning or selective telegraphy, which was not actually born till 1897. He is evidently impressed with the experiments both of Hertz and of myself, and he quotes from my *Philosophical Magazine* paper of August 1888 in confirmation and illustration of his prevision. For he says—after speaking of choosing wave-length with which to signal to specific people—" This is no dream of a visionary philosopher. All the requisites needed to bring it within the grasp of daily life are well within the possibility of discovery, and are so reasonably and clearly in the path of researches now being actually prosecuted in every capital of Europe, that we may any day expect to hear they have emerged from the realm of speculation into that of sober fact." And then he goes on—evidently referring to the experiments of D. E. Hughes, at which he must have been present—" Even now indeed telegraphy without wires is possible within a restricted radius of a few hundred yards, and some years ago I assisted at experiments where messages

D

were transmitted from one part of a house to another, without any intervening wire, by almost the identical means here described."

That article appeared in 1892, and was an inspiration of genius. Too little appreciation is felt to-day for the brilliant surmises and careful and conscientious observations of a great experimental worker like William Crookes ; and on some of it orthodox science still turns its weighty and respectable back.

In 1889 I had come across the effect of cohesion under electric impetus, and employed it to ring a bell under the stimulus of the overflow of a Leyden jar, as described in my paper to the Institution of Electrical Engineers in 1890. What I had specifically called a " coherer " was a needle-point arrangement, or the end of a spiral spring touching an aluminium plate ; which was and is extremely sensitive, but rather unmanageable. A crystal instead of an aluminium plate makes it all right, and it is now in constant use.

A less troublesome though similar method of detecting electric surgings was later discovered by Branly, in France, who was measuring the resistance of powdered metals, metallic smears on paper, and other finely divided substances. He found that the resistance of such smears varied capriciously, becoming much less when electric sparks were taken in the neighbourhood, but that a mechanical shock restored their high resistance. This was the same in essence as the coherer effect, and it was the origin of Branly's filings-tube, which, in one form or another, remained for some little time the standard method of detection, being employed by Popoff in Russia, Righi in Italy, Sir Henry Jackson, myself, and, in an improved form, Marconi. Branly's results were demonstrated in 1893 to the Physical Society of London by Dr. Dawson Turner, of Edinburgh, and thereafter optical experiments on Hertzian waves were rendered much easier than they had been.

CHAPTER IV

The Development of Radiotelegraphy

WITH a Branly's filings-tube I made many more experiments, developing the subject mainly on quasi-optical lines. And on the untimely death of Hertz I determined to raise a monument to his memory by a lecture at the Royal Institution on these experiments (Friday, June 1, 1894), which I styled " The Work of Hertz "—meaning that it was a direct outcome and development inspired by that work. I soon found that the title was misleading, so that in the next edition I changed it into ' The Work of Hertz and Some of his Successors," and afterwards changed it still further into " Signalling Across Space Without Wires " ; for that, of course, is what was being done all the time. The depression of a key in one place produced a perceptible signal in another— usually the deflexion of a spot of light—and, as I showed at Oxford, also in 1894 (employing a Thomson marine speaking galvanometer lent me by Alexander Muirhead), a momentary depression of the key would produce a short signal, a continued depression a long signal—

36

thus giving an equivalent for the dots and dashes of the Morse code—if the filings-tube were associated with an automatic tapper-back. One form of such tapper-back was then and there exhibited, a trembler or vibrator being mounted on the stand of a receiving filings-tube. This was afterwards improved, with Mr. E. E. Robinson's help, into a rotating steel wheel dipping into oiled mercury. Our aim was to get signals on tape, with a siphon recorder, and not be satisfied with mere telephonic detection. We succeeded ; but more rapid progress would have been made had we stuck to the telephone, as wiser people did.

My Royal Institution (1894) lecture was heard by Dr. Muirhead, who immediately conceived the desire to apply it to practical telegraphy. And when my lecture was published —as it was in the *Electrician*, with diagrams roughly depicting the apparatus shown, drawn (some of them) skilfully but not always quite correctly, by the then editor of the *Electrician*, Mr. W. H. Snell—it excited a good deal of interest, stimulating, to the best of my belief, Capt. (now Admiral Sir Henry) Jackson, Prof. Righi, and Admiral Popoff to their various experimental successes.

It was now possible to use them for roughly

and imperfectly transmitting signals in the
Morse code, either by the direct use of a Thomson marine speaking galvanometer or through
a relay by operating an ordinary Morse tape
instrument or a siphon recorder, which same
instrument, besides recording the signal, could
operate the " tapper-back." In August 1894,
I exhibited this method of signalling at the
British Association in Oxford. None of us
ever, so far as I know—unless it was Sir Henry
Jackson—actually applied these waves to anything that could be called practical telegraphy.
Nothing was as yet done outside the laboratory
and grounds, although Alexander Muirhead
was preparing to take the matter up from the
telegraphic point of view, clearly anticipating
the coming of practical wireless telegraphy
along these lines.

The first strenuous effort at actual telegraphy, though certainly the appliances were
then in an infantile stage, was made by Senatore
Marconi, who, having heard of the Hertz and
other experiments on waves from Prof. Righi,
of Bologna, worked away in his father's
garden in Italy, secured an introduction to
Sir William Preece, and came over in 1896 to
England.

Preece knew nothing of the experiments
that had been shown two years before in this

country : he was doubtless fully occupied with his official duties. But he was interested in the idea of wireless transmission, as he had made several attempts by means of electro-magnetic induction, and, indeed, this method, and the corresponding one by earth-tapping, can really be used, and, if there were no better method, might have been developed into something considerable. Preece realized, however, after a few demonstrations, that Marconi had in his possession a better method than ordinary induction between circuits. He regarded it as a complete novelty, and enthusiastically introduced Marconi and his device to the British public. The Post Office gave Government facilities, the attention of financiers was attracted, Prof. Fleming became scientific adviser and thereafter the matter became not only scientifically but also otherwise complicated. Progress was continually being made, distances were constantly being increased, and the transmission of Morse signals satisfactorily improved.

This long series of advances began in 1896. A most important early step onward was the achievement of accurate tuning by the introduction and utilization of adjustable inductance, with possible transformer reception, as recorded in my patent No. 11,875, of 1897, which was subsequently extended to twenty-one years by Lord Parker, and was then acquired by the Marconi Company from the Lodge-Muirhead Syndicate. This formed the necessary basis of all radio tuning. The validity of the patent was fully upheld by a long arbitration trial, under Lord Moulton, some years later.

A still further step in tuned telegraphy was made by Mr. Marconi in his famous Patent 7,777, of 1900, by combining tuned closed with tuned open circuits. He also devised a magnetic detector, somewhat on the lines of experiments made at Cambridge by Sir Ernest Rutherford. And then Fleming discovered his rectifying valve, and applied it to wireless, thus inaugurating more modern methods of detection. In the hands of Lee de Forest this became a triode, a magnifying or amplifying device of the utmost importance. By the use of valves, as everyone knows, detection is not limited to the discontinuities of Morse signalling, but we can detect all the continuous fluctuations which are employed in human speech. Consequently, not wireless telegraphy alone, but wireless telephony was born, after a short interval for further improvements. And gradually, in the hands of many experimenters—some of them military and naval, and some associated in one

way or another with the Marconi Company—
the apparatus became the singularly perfect
and efficient arrangement which is attracting
not only scientific but widespread popular
attention. The introduction of the vacuum valve as
rectifying detector by Prof. Fleming, and its
subsequent improvements, mark a great step
in the achievement of wireless telephony, and
has stimulated many amateurs to experimental
work of high value. It is needless to emphasize
the world-wide character of Mr. Marconi's subsequent developments, his discovery of the
power of ether waves to curve round the earth
to immense distances, his discovery also of the
adverse effect of sunshine, and the more recent
discovery that short waves can travel efficiently
to the Antipodes : the theory of which travel
by reason of peculiarities in the atmosphere has
been given both by Dr. Eccles and Sir Joseph
Larmor.

Looking back now at our old difficulties,
ever since the seventies of last century, the
ease and perfection of wireless telephony to-day
seems little short of miraculous. It just shows
what can be done by the combination of a great
number of workers all intent on securing
improved results. It is probable that many
who began as amateurs have contributed in

one form or another to this success. One hears of skilled amateurs who are able to talk to each other, when conditions are favourable and sunlight is absent, not only between America and England but between Britain and New Zealand.

I have also been told by correspondents in or near South Africa that they have been occasionally able to pick up wireless broadcast talks intended only for the British Isles. The rapidity of progress is wonderful, but it is quite unlikely that perfection has been reached. There is evidently no finality ; and we are probably still far below the summit of achievement.

CHAPTER V

Wireless Achievement and Anticipation

THE two things which the human race can effectively attend to, and achieve with some success, are *Locomotion* and *Communication*, both developed enormously and in an almost revolutionary manner during the nineteenth century ; and this development has continued during the early years of the twentieth century. Very few people still living can remember the introduction of railways into Britain. Some can remember the introduction of electric telegraphy ; many more, the beginnings of signalling by means of cables ; while electric means of transit, and wireless telegraphy are developments of our own time.

All electrical applications—from electric bells to the telephone, from the transmission of power by the dynamo to the latest messages across the Atlantic—represent the harnessing of the ether in the service of man. Whether a cable is used for the transmission is a mere detail. It is like using a speaking-tube instead of shouting across the street : air conveys the sound in both cases, but in one case it is guided, or so to speak focused, on a definite receiver ; in the other case it is broadcast. Electricity and magnetism and light are affairs of the ether primarily, though they are controlled, initiated, and directed by material appliances. But so far as mere transmission is concerned, matter is of no assistance, except that it can act as a guide like the walls of a speaking-tube. Electric force, magnetism, and light can go on equally well in a vacuum. To ether waves, matter is mainly an obstruction. Fortunately however, the air, in its normal state, has very little effect. It is essential to the conveyance of sound, but it takes not the slightest part in the conveyance of ether waves.

It is possible, however, to ionize the air, that is, to split it up into electrified particles—the positive and negative ingredients of which atoms are composed. Air thus acquires electrical properties, and is able to react upon the ether ; it becomes a conductor, though a poor one. Such air, like any other electric conductor, is partially opaque to ether waves, and, like other opaque materials, it can either obstruct those waves, absorbing them and turning them first into electrical currents and then into heat, or it can reflect them, somewhat as a mirror reflects lights.

Many causes are capable of ionizing air.

Radioactive substances do it, though they themselves are a recent discovery. But the sun is a radioactive substance on a large scale, and undoubtedly some ionization of the atmosphere is due to solar radiation. There are other causes, such as the splashing and spraying of water and the breaking of water-drops, which by some eminent physicists are considered to be capable of accounting for most of the electrification of clouds, and the consequent occurrence of thunder-storms.

Electric discharges in the atmosphere on a small scale are very frequent ; and they are known in radiotelegraphy as " atmospherics." They are of no assistance, and are a nuisance which ought to be eliminated. One of the problems to be solved is how best to eliminate their disturbing effect on the reception of messages. Moreover, when ionized air exists extensively between a sending and a receiving station, it acts as a partial screen and renders communication difficult, in the same sort of way that a light fog or mist causes indistinct visibility.

Ionization is not wholly obstructive. An ionized layer of air might assist transmission by its refracting power. It might, for instance, cause the rays to move in a curved instead of a straight path. Such a helping layer is believed to exist in the upper regions of the atmosphere, for it is those upper regions which receive and consume much of the specially active rays from the sun. Waves generated at a sending station are, therefore, liable to be curved round the earth by this ionized layer, when it is placid and not too corrugated, somewhat like a mirage, or roughly as a whispering gallery acts in the case of sound.

Water also is a conductor, which can still more efficiently reflect the waves, and thus we live between two layers—a " floor " of water or damp soil, and a " ceiling " of ionized air—so that ether waves cannot easily get out and travel across empty space, which they are so well qualified to do. They are enclosed, as it were, in a space of two dimensions—a most fortunate circumstance, without which wireless telegraphy at a great distance would be impossible. Rays travelling in straight lines, like lighthouse beams, could not possibly travel, say, from London to New York, whatever their intensity. They would go far over the top of a receiving station; even at a distance of only a few hundred miles.

Summarizing what I have said in a previous chapter :—The discovery of electric waves was made in the latter half of the last century by that tremendous mathematical genius Clerk

Maxwell, on the purely theoretical side. After twenty years, Hertz showed how to produce them practically, and what was more, how to detect them at a distance, in an elementary and purely laboratory fashion. Further improvements in detecting appliances were soon devised by many people, and in due time they were made amenable to practical and commercial uses by the energy and enterprise of Senatore Marconi and his co-workers.

To a public ignorant of the work of Clerk Maxwell and Hertz, this application came as a great surprise, and seemed very novel and mysterious. To physicists it did not seem so : it was a natural application of what was known. But when Senatore Marconi found experimentally that the waves would actually curve around the earth and reach the American continent, physicists were surprised. It was an important discovery; and a mathematician, Mr. Oliver Heaviside, began to show how an ionized layer of air in the upper regions must be operative, and could explain it.

Tuning and selective telegraphy were realized by the proper use of self-induction, as set forth in the fundamental Lodge patent of May, 1897.

Then came a method of detection far superior to any that had previously been used, namely the vacuum valves of Prof. Fleming, improved, as they soon were, into their present form by Dr. Lee de Forest of America. In these valves the actual electrical particles, the electrons, were employed as the detecting agency, and proved themselves far more perfect than any material mechanism could be. They responded instantaneously to every fluctuation, so that it became possible to transmit, not Morse signals only, but microphonic or telephonic speech.

For some time it seemed as if speech could only be transmitted over moderate distances. But now, through the energy and enthusiasm and inventive genius of a great number of workers in all parts of the world, but especially in England and America, it has been found possible to hear the human voice across the Atlantic. Not that the voice travels any farther than it did before, any more than it travels along a telephone wire : the voice generates electric waves, with all its peculiarities accurately represented in those waves, and when those waves are collected by a distant aerial, the electrons in the receiving valve respond with precision to all the fluctuations, and enable a telephone to reproduce the speech and the tones of voice of the distant speaker. The achievement of speech across the Atlantic in this indirect way is certainly a marvellous one, which excites the admiration and to some extent the astonishment even of physicists. Nor is this likely to be the limit. The waves that have begun to curl round the earth can go on even to the Antipodes, and in a short time it is likely that the human voice in this way can reach Australia and New Zealand. Thus humanity will be welded together in a manner more intimate than ever before, and the beauty and simplicity of the arrangements, the comparative ease with which the result is effected, is very surprising.

It used to be thought by the early experimenters that to get waves to travel effectively over enormous distances, the apparatus used must be large and powerful and the waves very long. Long waves can usually get through obstacles which would stop short ones. Why ? Because in going through a slightly opaque medium a certain percentage of energy is wiped out at every swing. The crest of each wave will be slightly weaker than its predecessor. Therefore, if in a given distance, say 100 miles, there were twenty crests—which would mean that the waves were 5 miles long—there would be a chance of a sufficient portion getting through, even though each wave were 1 per cent. weaker than the one behind it. But if the waves were only a quarter mile long, there

E

would be 400 such crests in the 100-mile distance, and the proportion of energy which got through would be very slight. If the waves were each only 100 yards long, the oscillations in the given distance would be so numerous that no trace could be detected, unless the opacity were insignificant. Hence, it appeared that long waves had the advantage.

To the physicist it always seemed that short waves ought to do better, if space were as reasonably transparent as one might expect it to be ; that is, when the air is hardly ionized at all, which is a condition to be expected in the absence of light. Short-wave radiation is far more intense than long ; a much greater conversion of energy into radiation takes place. An engineering alternator hardly radiates at all ; its waves are far too long. A simple dumb-bell set oscillating may have but little energy, but whatever it has it radiates completely. There must, therefore, be a compromise for powerful signalling by waves. It is now found that, at any rate during the night, short waves are quite efficient, and the great size of sending and receiving stations will probably, in due time, be found unnecessary. A short-wave or small station is just as energetic as a big one, within limits. It possesses less energy, but it radiates a larger proportion. For the true wave starts not at the actual radiator, but about a quarter wave-length distant from it. Hence, the shorter the wave, the nearer, and therefore the more energetic, is the place from which it starts. A radiator no bigger than a dumb-bell can emit waves of 100 horse-power. This I published as long ago as 1890. A very large radiator under the same conditions is no more intense, though it is true the emission would last longer. That would depend on its capacity. And what is true of the emitter is also true of the receiver. Hence, recent experiments have redirected attention to the advantages of short-wave transmission.

Moreover, short waves are much more amenable to discipline. They can be projected by parabolic mirrors of reasonable size, as Hertz showed long ago, in 1888 ; that is, they can be directed, as light waves are directed from a lighthouse, so as to economize them and concentrate them in any required direction. There can be little doubt that this power of focusing and directing waves will be applied more and more, so that except for broadcasting purposes it will not be necessary to send out waves in every direction, at random. Senatore Marconi is beginning to apply this on a large scale.

Another improvement which is to be ex-

pected is the attainment of greater power of
control over distant things like aeroplanes and
steamers, or other self-propelled floating bodies.
The rudders of such machines can be actuated
by people on them, but they may also be actuated wirelessly by people at a distance, so that
an operator at a sending station, manipulating
his keys, may guide a distant floating body to
any desired destination, so long as he can see
what it is doing and adapt his control accordingly. An aeroplane is not so easy to adapt
as a floating body, because it has another degree
of freedom. It can move up and down, as
well as right and left. To control it perfectly
is, therefore, not so easy. But none of these
things is easy. Difficulties are things to overcome, and the ingenuity of those who are
working at the subject is more than competent
to deal with a difficulty such as this. It is
interesting to find, moreover, that the old-
fashioned coherer, employed as a detector,
seems especially useful in these distant-control
experiments, as has been demonstrated recently
by Major Phillips.

What other developments are to be expected ? Unfortunately, a certain amount of
energy in the present state of civilization is
directed to the opportunities for doing damage ;
that is, directing things for deleterious pur-

poses. If people wish to do those things, no doubt they will always be able to find ways and means for so doing. It has been surmised that aeroplanes can be stopped in mid-air. Well, as Hertz found long ago—and before him both Joseph Henry and Elihu Thomson— ether waves are powerful enough to generate little sparks in metal conductors ; and as the explosions of oil vapour in a motor are regulated by little sparks, it seems quite likely that such sparks can be generated at wrong times by the action of waves generated at a distance. If so, the engine may be brought to a standstill by the generation of unexplainable engine trouble. Such disturbances can be guarded against, when foreseen, by the proper use of a metallic screen or enclosure, because metals are opaque to the waves, and will ward off or reflect them harmlessly. A screen to be efficient must be complete, in the sense that it must have no bad joints or unclosed chinks. Round apertures do not let waves in, but elongated chinks do. A bird-cage complete with metallic floor is an efficient screen, but any insulated conductor penetrating its meshes would guide waves in. Electric leads and gas-pipes are liable to turn an otherwise protected enclosure inside out.

Contrivances for doing damage are danger-

ous until the antidote is found. There always
is an antidote, but meanwhile much damage
may be done. It is lamentable that the
ingenuity of man is thus capable of being
misdirected. Other things can be suggested
of a damaging character, though it is hateful
to dwell upon them, and it is not a subject on
which I am any authority.

However great have been our improvements
in locomotion and communication during the
past hundred years, that is but a small period ;
and who can say what will be accomplished in
the next hundred years ? Material progress,
however, is not everything. And if there were
any signs of our getting to the end of our tether
in this respect—which there are not at present
—there would be no reason to lament.

Locomotion is purely a physical thing, but
communication, whether by speech, writing or
telegraphy, is not solely a physical thing. It
is a psychical thing, too. There were those in
the sixties and seventies of last century who
lamented that many of the messages sent
through the recently achieved Atlantic cable
were either deleterious or rubbishy. It is no
use enlarging our powers of communication if
we have nothing worth while to say. The
moral and spiritual development of mankind
ought to keep pace with its material achieve-

ments. And if they do not, it is possible to regard even those achievements with gloom and apprehension. That, however, would show a lack of faith. The real progress of humanity is necessarily slow, while material achievements may be rapid : it rests with ourselves whether or not one can keep pace with the other. There should be no feeling of supine self-satisfaction in what has been done, but a girding up of our energies to see that the progress is not too lopsided and unbalanced, and to contrive that the reign of good shall keep pace with the reign of power.

CHAPTER VI

Vast Range of Ether Vibrations

IT is of interest to call attention to the fact that what is called the spectrum—that is to say, the known range of vibrations in the ether—is now nearly complete. By different methods it is now possible to detect and obtain rates of vibration ranging from those of quite low frequency, expressed by such small figures as 1, or even a fraction, per second, up to those which are so immensely rapid as to be almost uncountable.

To deal with the slow ones first, the capacity of a farad joined to an inductance of a henry would have an oscillation period of six seconds, which is about the same as the oscillation period of a charge upon the sun. On the earth a charge would complete an oscillation in the seventeenth part of a second. A microfarad connected to a henry of inductance would oscillate a thousand times in six seconds, and so generate a feeble wave 1,800 kilometres long. To get anything like strong radiation we must quicken the rate of vibration, and shorten the wave ; but a very practical wave,

1,800 metres long, with a frequency of vibration about 170,000 per second, can be got by coupling a millimicrofarad, or nine metres capacity, to a millihenry, or 10 kilometres inductance. It is still easier to get waves of great intensity only a few metres long. A wave of 300 metres has an oscillation frequency of a million per second ; and with care and precaution these so-called wireless waves can be shortened in the laboratory down to something like a centimetre, which would correspond to thirty thousand million vibrations per second.

So already the electrical rates of vibration are getting considerable, but still nothing like those which we have learned to associate with ordinary light.

The range of luminous vibrations, that is, those which can affect the eye, and therefore are popularly called light, is, as is well known, limited to "an octave," ranging from about 400 to 800 millions of millions per second. But below the visible range we have the infra red, sometimes called "heat" waves, extending downwards without anything but an experimental limit, until they almost reach a range of extremely high electrical vibrations, such as those above mentioned, rising up to meet them. Electrical vibrations go on extending downwards, through the great range of wireless waves with frequencies of anything from a million to, say, ten thousand per second, to the slow oscillation of large capacities joined to great inductances, such as one might have in a transformer station, or with alternating dynamos; it being understood that the radiation from these slower things is insignificant, and that the radiating power increases (other things being equal) with the fourth power of the frequency of the vibration.

At the other end of the scale, above the visible range, we have ultra-violet radiation, extending into the photographic region without obvious limit. There has been a practical limit until lately, but now the range has been extended, by photo-electric devices, until it overtakes and begins to overlap the soft X-rays. And these rise, through ordinary X-rays of excessively high frequency, up to the gamma rays emitted by radium, which at present constitute the highest terrestrially known rate of vibration, some hundreds of thousands of millions of millions per second.

It is possible that in the sun, or especially in the interior of some of the hotter stars, there may be rates of vibration even higher than that, due to the disintegration of atoms and the excessive temperatures which would be there encountered.

All these higher rates of vibration would be very deleterious to us ; but fortunately they are easily stopped by a thin layer of matter, so that from the stars they hardly emerge, while those from the sun are screened from us by the earth's atmosphere. We only encounter a few of them when we ascend to great heights ; and then we do experience their blistering effect.

It used to be taught that the solar rays consisted of three different constituents merged into one—thermal, luminous, and actinic rays— though such an opinion was never seriously held by experts. The truth is that the waves differ in nothing but wave-length or rapidity of vibration, but that the short waves find something in atoms to respond to them, and thereby excite vision and chemical action, while all waves generate heat in proportion to their energy when absorbed, long waves being usually the most energetic.

It is beginning to seem probable now that the earth is kept warm by the absorbing power of a layer of ozone in the upper regions of the atmosphere, which has the power of stopping a good deal of radiation and of becoming warmed by it ; thus constituting a sort of blanket, and preventing us from ever feeling the full intensity of the dread cold of space, which must be a close approximation to absolute zero. When the sky is clear and the sun is set, we do feel some traces of this cold, and that is what gives us our hard frosts. But for the most part, the earth as a whole is mercifully screened from the more violent ranges of temperature : otherwise life could not have persisted and attained the approach to perfection which in the course of millions of centuries it has attained. Presumably there is some kind of similar provision on most of the other planets ; and accordingly it appears probable that life of some kind—though not necessarily human life—would be found on them also.

By the planets here mentioned we mean the planets of the solar system, the only planets of which we have anything like adequate knowledge. What may be happening on the innumerable other planets which may be circulating round the infinitude of stars in space, we have at present no conception. But the universe is so majestic, and its possibilities so immense, that no one with any wisdom would venture to put a limit to the possibilities and variety of existence.

THE EARTH A HEAVENLY BODY

We seem to have travelled far afield from the more or less practical considerations with which we began. But now that we are beginning to deal in an intelligible and practical manner with the ether—that universal medium which unites all the worlds—no one can say what may be the ultimate outcome. The ether has already brought us much information as to the chemical constitution and other details of what are called the "Heavenly Bodies"— though it should always be remembered that the earth is one of them, though a small one, yet just as much a heavenly body as the others, difficult as it may be occasionally to believe it, or to reconcile that fact with some of the doings of humanity The ether, I say, has already brought us so much information about the heavenly bodies, that it will surely in the progress of science bring us more ; and so in due time we may receive quite unexpected information about them.

For science is as yet in its infancy. Our methods of exploration are continually enlarging ; we have already found that we are not isolated and disconnected from the rest of the universe as we formerly believed, and as in olden times, for all practical purposes and by the methods of science, we were. Though it should always be remembered and admitted that, by methods other than those of science, men have always believed themselves to be in touch—at first in awe-stricken but afterwards reverent and even affectionate touch—with a higher order of existence. Things half known, and but dimly glimpsed by the ancients, may in process of time become known to us, through the accumulation and passing on of laboriously acquired knowledge.

And just as the higher and lower regions of the spectrum have gradually united, so that some approach to continuity is established through the whole range, so it may be hoped, and even confidently expected, that in the long run the regions of knowledge and of faith will approach each other by gradual extension, and merge into a comprehensive unity.

CHAPTER VII

The Transmission of Wireless Waves

THERE seems to be a good deal of misunderstanding as to how electric waves are propagated from an aerial, not only as regards the distance travelled and the way in which they get round the curvature of the earth, but as to their actual mode of propagation, and as to what process is going on in the ether which is able to advance with the velocity of light. For electric waves are not only electric, they are electro-magnetic. That is to say, they have an electric component which is detected at a receiving station by an elongated, or linear, conductor ; and they have a magnetic component which is detected by a closed loop or coil of wire. These are the two kinds of aerials in common use, the elevated wire and the closed loop. One responds to the electric, the other to the magnetic oscillation ; and it is pretty well known that these two oscillations are at right angles to each other, and that it is most efficient to have the electric one vertical and the magnetic one horizontal. It may also be known that they have equal energies; their energies are necessarily equal, so that any weakening of one equally weakens the other. The whole progress of the wave depends on the co-existence of these two forms of energy, the electric and the magnetic : and if one stops, they both stop. If one is reversed, the other must be reversed too if the propagation is to continue in the same direction. If one is reversed without the other, the wave goes backwards. And if at any place the one exists alone, the wave stops, and at that place you have either an electric phenomenon or a magnetic phenomenon, but not both.

The consequence of all this is that the electric and magnetic disturbances must be coincident in position; one cannot lag behind the other in a true wave. Whenever one is at a maximum, the other must be at a maximum, which is expressed by saying that they must be in the same phase, as a condition of the progress of the wave.

Yet it is often taught that one is a quarter period behind the other, like the piston and slide valve of an engine ; so that when one is at the extremity of its swing, the other is in mid course ; and that the energy oscillates from one form to the other, being alternately kinetic and static. For magnetism is due to current or kinetic energy, while electrification is due to static or potential energy; and in ordinary cases they do not co-exist. You may have an electric current, or you may have a charged body. Wherever you have both, you have oscillations and the generation of waves.

But the curious thing is that at the generator the energy really does oscillate from the static to the kinetic form, and back again. Consider an ordinary aerial, with a capacity area above and below, and a coil in the middle between them. At one instant the upper area is charged positively, the lower area negatively, and there is no current in the coil. At the next instant, separated from the first by a quarter period, the current in the coil is a maximum, and neither area is charged at all. In half a period from the start, the current has stopped again, having piled up its momentum in the two areas in the form of a reverse charge, the lower being now positive, and the upper negative. This sets up an elastic strain which recoils back again, generating an inverse current in the coil; which current reaches a maximum, and then expends its energy in recharging the areas in the original way. And so on periodically. The process just recorded is a complete period, and occupies, of course, a very minute fraction of a second, even with the biggest aerials.

F

Hence at the emitting station the electric and magnetic disturbances are not in phase. One lags a quarter period behind the other, just like the slide valve and piston of an engine. A little way off in the ether the conditions have become different. At a distance of about a quarter wave-length the electric and magnetic disturbances have caught each other up, and got into phase. Within that quarter wave-length they are not in phase ; and, accordingly, the energy in that space oscillates to and fro, alternately travelling outwards and travelling backwards, from and to the source—a pulsation in the ether—and no true wave is broken off or emitted within the first quarter wave-length. But at a certain distance, which was calculated by the great discoverer, Heinrich Hertz, in the light of Clerk Maxwell's theory, some of the energy is flicked off at every oscillation. At that distance the two etherial disturbances have got into phase. They are coincident with each other, and when that happens the only way in which they can co-exist is to fly along with the velocity of light ; which accordingly they continue to do, until their energy is somehow absorbed or dissipated by conductors. Hertz gave diagrams of the whole process, according to Maxwell's principles, about the year 1890, and thoroughly understood it.

That is why an ordinary alternating dynamo of commercial frequency emits no appreciable waves. The place whence waves would start is a quarter wave-length away. And if the oscillations are a hundred a second, the wave-length is 3,000 kilometres, or say 2,000 miles, so that the quarter wave-length is 500 miles. And the waves from an alternator of 100 a second in the south of England would not begin till about the distance of Aberdeen : that is to say, practically they would not begin at all, though theoretically it is true that every alternator must emit waves of infinitesimal strength. But the waves only become strong and important when the frequency of oscillation is very great ; and the higher the frequency, that is to say, the shorter the wave-length, the greater is the proportion of energy emitted in radiation. The advantage of long wave-length is not that more energy is emitted for a given horse-power of the sending station, but that the waves are better qualified to overcome obstacles, and to travel to a great distance without so much loss.

That is a digression. What I want to say, further, is that the process of wave-transmission, which has been described and worked out for electro-magnetic waves, is essentially true of all waves. The kinetic and static energies are not oscillating from one form to the other, but are coincident and travelling together. Prof. Howe has recently pointed out that it is true even of sound waves. At the place of greatest compression or rarefaction we might have thought that the particles would be stationary. So they are in an oscillating column, like that in an organ pipe. So they are in any source of sound, but not so a little distance away : not so in a sound wave, as distinct from the alternating pulse which generates a sound wave.

When we study the phenomenon in a true wave we find that the particles in a condensation, or greatest compression, have likewise their greatest speed. They are travelling full-speed forward, while in a rarefaction they are travelling full-speed backward. The static and the kinetic energies agree in position, just like the electric and magnetic. It is at the intermediate parts of the wave that we find them both momentarily at zero. The particles are stationary at the places where the air is of average density, not in a compression or rarefaction. Hence the theory is very general, and those models which have been constructed to illustrate the propagation of waves, and to show the lag of one form of energy on the other, are erroneous. They only apply to the oscillator, not to the waves. So-called stationary waves, the result of reflexion, are essentially akin to an oscillator. True waves must advance.

The fact that the true wave only starts a quarter wave-length away from the oscillator is very instructive. It applies even in the case of light, although in that case the oscillator is of ultramicroscopic dimensions, and the frequency hundreds of millions of millions per second ; so that the following-out of the process in detail might seem impossible. But it was not impossible to the great mathematician, Sir George Stokes, who in his work on fluorescence arrived at the conclusion that the quarter-wave lag or difference of phase at the start must be compensated or neutralized, so that it became obliterated in the true wave.

It is in many respects the same even with waves on the surface of water. The particles of water are moving forward on the crests, and are moving backwards in the hollows. They are moving only up and down at the position of mean level. If you watch sea-waves travelling along in deep water, you will not at first notice the motion forward of the particles at the top of the crest, since straws and ripples on the surface go backward relative to the wave as it advances. But that only means that the water particles which are moving forward are not moving at anything like the speed of the wave itself. The wave is going much faster than the particles, and hence overtakes them, and slides under them. The speed of the water particles varies with the amplitude or magnitude of the disturbance. The speed of the wave does not depend on that at all, but only on the wave-length, that is, on the distance from crest to crest ; whether the wave is a mere inequality of the surface, or whether it rises 20 or 30 ft. The velocity of the wave —the speed with which the crest itself advances —depends not at all on the height or intensity of the wave ; but it does, in the case of 'a water wave, depend on wave-length, i.e. on the distance separating successive crests. In fact, in deep water the velocity of wave-progress varies with the square root of wave-length, for big waves. For ripples the law is different.

SOUND AND LIGHT SIMPLER THAN WATER WAVES

All these things are complications which we do not find in the ether, nor even in the air. The speed of sound depends on the conveying material only, not on loudness, nor even on wave-length or pitch. Sir Isaac Newton realized that, for he pointed out that a band heard at a distance could not possibly sound like music unless every note, loud or soft, high or low, had one and the same rate of travel. So it is also in the case of light and wireless waves. They all travel through the ether at one identical pace, whether they be a hundred miles long, or the millionth of an inch short. Also whether they be bright like sunlight near the sun, or dim like a rushlight or a glow-worm. In this respect, therefore, ether and air waves differ from visible waves on the surface of the water. But all waves agree in this, that the potential and kinetic energies—that is, the displacements and the velocities—are concurrent in phase, rising to a maximum and falling to a minimum together. This is a peculiar condition, destructive of equilibrium, and it can only be satisfied by the wave advancing through the medium at its own proper pace—a pace which in wireless waves is determined by the mutual reaction of the electric and the magnetic components, in accordance with what is called Poynting's theorem.

A receiver acts by obliterating some of the electric component, and thereby stops a portion of the wave. This it does either directly as by a linear aerial, or inductively, as by a loop aerial. The energy of such portion of the wave as effectively encounters the aerial is abstracted and utilized for the signal, some fraction of it degenerating into heat. The rest of the wave goes on.

So to sum up. The electric and magnetic components of a wireless or electro-magnetic wave are at right-angles to each other, and are equal in energy and coincident in phase, so that both reach a maximum, a minimum, or a zero, together. There is no lag of one behind the other, such as occurs naturally in all our emitting and receiving instruments. And the only way in which this curious unstable condition of things can be sustained is for them both to advance forward with the velocity of light. And that is just what they do. The oscillator is stationary, true, but then the two disturbances there are not in phase. One is a quarter period behind the other, as one would expect : then the energy mainly pulsates, first out, then in, and is not all lost by radiation. The only part lost by radiation is that which has got a quarter wave-length away, where the one disturbance has caught up the other, and where the energy—that which is used in wireless telegraphy—is flicked off into space.

CHAPTER VIII

Wave Peculiarities

WAVES are transmitted by a combination of momentum and elastic recoil; and, in order to convey them, the medium must have the two corresponding properties, viz. something corresponding to inertia or massiveness and something corresponding to elasticity. There is a displacement from the mean position, with a tendency to spring back, this displacement being either material or electric, according to circumstances ; and there is a rushing past the mean position with a momentum which overshoots the mark, and carries the particles into a region of recoil, propelling them against the electric force, which in due course drives them back again. It is the elastic force which generates the momentum ; and it is the momentum which piles up an opposite elastic force. It is by the interaction of these two properties that oscillations are maintained, and it is by possession of these two properties that the medium is able to pick up the oscillations and transmit waves.

In the case of mechanical waves the momentum is straightforward mechanical momentum, due to the inertia of matter ; in electric waves the momentum is magnetic momentum, and is due to the inertia of ether ; just as the elastic recoil is due to the electrical rigidity or elasticity of ether—the term " rigidity " being a technical or quantitive one, not at all implying infinite rigidity.

The reaction of the two properties is easy enough to follow in the oscillator, or source of the waves, where the two are in different phases. It is rather less easy to follow in the wave itself, where they are both in the same phase. But it may help if we consider the simplest case, viz. that of sound. As the wave advances the particles are simultaneously thrown forward and nearer together, so as to make a condensation. If at the maximum condensation they came to rest, they would rebound, and the wave would go backwards. That is exactly what happens when sound encounters a wall or other obstacle and is reflected, giving rise to what are called " stationary waves," with nodes and loops in definite places. But in a progressive wave the particles are thrown forward into the condensation, and the condensation moves on by reason of the momentum of the particles, and so continually advances into new regions, spreading the disturbance at a steady advancing pace. The particles ultimately come to

rest, and go back again ; but when they do
that the rarefaction is beginning. And so this
rarefaction, or pull on the medium, travels forward likewise ; though it requires a little more
effort to follow the details of the advancing
rarefaction, as compared with the more easily
grasped details of an advancing condensation ;
since in a rarefaction the particles of the
medium are moving in a direction opposite to
that of the advancing wave—which at first
seems a little confusing.

To follow out the details in the electro-
magnetic case is less easy, because we are less
familiar with the intimate nature of electricity
and magnetism. We know that magnetism is
the result of an electric current (or, rather,
that the two are different names for the same
thing), and that it simulates mechanical inertia
and momentum. We also know that electric
displacement is another name for electric charge,
and that it simulates an elastic spring-back,
which is demonstrated by an electric discharge.
The term " charge " is only applied to a conductor: the more general term is electric displacement, because that applies to an insulator
too. But, as the process is not a mechanical
one, the only way we can safely and completely
follow it is by the use of the vitally important
equations of Clerk Maxwell, which contain the

whole theory embedded in their intricate and illuminated recesses.

But now, to understand the phenomenon of wave transmission more thoroughly, we ought to ask whether there is any kind of wave which does not obey those laws—which has not both momentum and elastic recoil, which does not require those properties in the medium, and which does not advance at a definite pace.

The answer is that there are such waves— if they can be called waves—waves of diffusion, transmitted from an alternating source. It may be a source of alternating temperature, for instance, like the summer and winter temperatures applied to the crust of the earth. The result is that periodic waves of temperature sink into the crust, and succeed one another at regular intervals ; so that by penetrating a sufficient depth you will find a trace of last summer's heat, and by penetrating deeper you will find a trace of the preceding winter's cold, and so on, though it is true that these traces tend to smooth themselves out rapidly, and become before long difficult to recognize. But if we ask at what rate these waves travel we can give no answer, at least no clear and simple answer, as we can in the case of true waves with momentum. For the peculiarity of these heat waves is that they have no momentum.

They travel according to a different law ; and the time taken for any given portion of heat to reach a certain depth will depend upon how sensitive the detecting instrument can be.

We may illustrate the matter by taking a long rod of metal, packed in cotton wool or some insulating material to prevent loss of heat from the surface, and then put a thermometer at one end (which might be a sensitive thermopile), and to the other end apply heat and cold alternately ; for instance, first a flame and then a douche of liquid air. The thermometer at the far end will exhibit alternations of temperature. There will be a lag, perhaps a very considerable lag, before it feels the first heat wave. The cold wave may be well on its way, and another heat wave too, before it responds ; but in due time it will go up and down in response to the succession of impulses imparted to the other end. But if we ask how soon it will feel those impulses after they started, we must realize that it is merely a question of how sensitive it is. If it can only respond to a Fahrenheit degree or two, it will be very sluggish ; but if it can feel the millionth of a degree, it will be fairly quick.

This is the actual problem which had to be solved in connexion with the first Atlantic cable, and Lord Kelvin solved it completely.

He perceived that the electric capacity of the cable would make the laws of electric propagation correspond exactly with the laws of the flow of heat—which had been worked out by Fourier. So he gave the theory of the cable, treating it as a long conductor to one end of which positive and negative electrification was alternately supplied, a detecting instrument being at the other end. And he realized that if signalling was to be at all rapid, this detecting instrument must be of surpassing sensitiveness. He knew that there was no true velocity of propagation in a cable possessing only resistance and capacity ; he knew that the waves were waves of diffusion, having no definite speed, and that the rapidity of their detection must depend on the sensitiveness of the receiving instrument. Hence his mirror-galvanometer, and then his siphon recorder. He knew further that violent applications of electricity to a cable were unnecessary and troublesome, as well as dangerous ; that they put into the conductor disturbances which would have to leak out ; and that within limits the feebler the signals were, the better. Sharp momentary curbed signals ought to be sent, and though at the far end they arrive washed out by diffusion, yet a sufficiently sensitive instrument can detect them without more than a fraction of a second delay, and with a reasonable amount of sharpness.

But this theory after all was not complete, was not quite complete. It left out of account self-induction ; that is to say, it treated the electric waves as if they were really like heat waves, without momentum. As a matter of fact, the heat wave theory is only an approximation or an analogy. Electric waves must have momentum, since every electric current has a magnetic field round it. The magnetic-inertia effect was omitted in Kelvin's theory. If thought of at all, it was thought to be insignificant, wiped out as it were by the capacity and the resistance ; just as sound waves trying to pass through a haystack would have their momentum wiped out by friction, and would be stopped. Or, a better analogy, as light waves are stopped when they encounter any black material. The propagation of heat waves through a good conductor, the propagation of light waves through a bad conductor, and the propagation of electric oscillations through a long submarine cable, all follow the same law —the law of diffusion—the law according to which a coloured solution like sulphate of copper is conveyed along a tube of water ; the law on which a particle of salt dropped into a bucket of water is ultimately found to permeate every part, even if the water is not stirred but continues stagnant.

But whereas in the chemical case and the heat case the absence of momentum is complete, in the electro-magnetic cases it is only approximate. And when electro-magnetic momentum is taken into account, the law is somewhat modified ; an element of true wave makes its appearance, and there is a true velocity of propagation, though it may be only for an insignificant part of the wave. The head of the wave, however, does advance with the velocity of light, though it may be a head so small as to be undetectable.

But it may be strengthened, and the way to strengthen it is to increase the magnetic momentum, that is to say, to increase the self-induction, as might be done by coiling the wire upon itself, or by surrounding the wire with a sheath of iron. In that way the diffusion effect can be minimized and the wave effect strengthened, with immense gain in rapidity and clearness of signalling. The complete Theory of the Cable exhibits all this, and enables quantitive calculations to be made. And that complete theory the world owes to Mr. Oliver Heaviside, whose brilliant investigations were not recognized at the time by those in high telegraphic authority. He seemed to be merely complicating matters, and not introducing anything practical. So no practical improvements resulted, until ultimately Lord Kelvin himself perceived the merit of Heaviside's extended theory : and Silvanus Thompson in England, and Dr. Pupin in America, advocated the introduction of special self-inductance coils into a cable, especially one that was to be used for telephonic purposes.

Undoubtedly the construction of such electrically improved cables was a real difficulty ; and the men who had to lay them naturally shied at additional complications to a problem which in its early form was difficult enough. Moreover, the Kelvin instruments had enabled so much speed and certainty to be obtained in ordinary Morse signalling through a cable, that there was no great stimulus to a revolutionary improvement.

But when it came to telephony, the case was different. Telephonic speech was impossible if all the consonants and vowels were diffused and run together in a vague indistinguishable mass, with no genuine velocity of propagation ; so that as it were the stronger waves arrived first, and the feebler ones never arrived at all, and the peculiarities and harmonics of a tone (on which speech depends) were blotted out in transit. Speech through a submarine cable a

G

hundred miles long is impossible, unless extra self-induction is introduced. Even forty miles long is difficult, and the difficulty increases with the square of the length. But if sufficient self-inductance is introduced, so as once more to gain magnetic momentum, true wave propagation is restored ; the signals travel with a definite speed, independent of intensity and of pitch, and arrive with their features fairly intact, weakened no doubt, but not disfigured. The deleterious effect of capacity combined with resistance remains, and there is some diffusion as well. Speech through a cable is never so easy as through a land line, or through the unconfined ether of space. But it becomes possible ; and in the light of Heaviside's theory, the possibility is clearly intelligible.

The great advantage of radio-transmission is that these diffusion difficulties do not occur, true waves are propagated at one definite speed, every feature is retained ; and so speech through free ether has no limit of distance.

CHAPTER IX

On the General Theory of Ether Waves

IN Chapters II and III I have gone over the early past history chiefly in order to recall to your attention the extraordinarily brilliant work of Clerk Maxwell. It was not of a kind to get into the newspapers or to be understanded of the people. But in the students of those days—I mean the students of mathematical physics—it aroused the utmost enthusiasm ; and it is well that this mighty work in pure science should not be forgotten and overwhelmed in the mass of practical detail and exuberant invention which is a characteristic of the additional developments in our own day. "All can raise the flower now, for all have got the seed." But it is well occasionally to think of what we owe to the sowing of that seed, as sprinkled in algebraical symbols on the pages of Clerk Maxwell's papers, and incorporated, some of them, in his great book.

Since that time Sir Joseph Larmor has shown that Clerk Maxwell's equations contain the whole theory of radiation, due to the relation between electric charge and the ether of space. He has also shown that radiation is always

excited to a known extent whenever an electron is accelerated, and that without electric
acceleration there is no radiation. I must just
give Larmor's expression for the radiation of
energy, because it is very fundamental. It is :—

$$\dot{W} = \frac{2\mu}{3c} (e\dot{u})^2$$

This gives the energy radiated per second from
an accelerated electron, where e is the charge
of the electron ; c is the velocity of light ; $\dot{u}$
is the mechanical acceleration ; and μ the magnetic constant of the ether.

Why is that expression so important ? It
is because it is exceedingly general and comprehensive. Radiation of so many kinds is
known—X-rays, wireless telegraphy, light,
radiant heat—all these obey that law—there
is no radiation if an electron is not accelerated.
There are three possible states : an electric
charge stationary is one of them, an electric
charge at steady speed is another, an electric charge with varying speed, or acceleration, is another. You have the three great
departments of the science. Static, kinetic,
oscillatory. In the first the charge is at rest,
i.e. ordinary static electricity. In the second
the charge is in motion—steady motion—that
is, current electricity and magnetism. Directly

a charge is put in motion you get magnetic lines surrounding the path of the charge. That is all that magnetism is: you don't get any magnetism unless there is a moving charge. In the third, the charge is accelerated, and that is light, because in any oscillation there must be acceleration. The connexions between these three groups are indicated in the table :—

Charge at Rest	Charge in Motion	Charge in Accelerated Motion
|	|	|
Static	Kinetic	Oscillatory
|	|	|
Electricity	Magnetism	Optics

Wherever you have acceleration, and therefore whenever you have oscillation, you must get radiation to an amount which depends upon the frequency in the way I have been explaining, and the amount of energy radiated depends upon the square of the acceleration—i.e. upon $\ddot{u}^2$ ($\ddot{u}$ being the first time-differential of the velocity u). If an electron is moving steadily it does not radiate at all, it does not lose any energy—a magnet loses no energy. When its velocity changes, it radiates energy depending on the square of the acceleration. It must, however, be admitted that the acceleration to be effective involves a change of speed: steady circular motion does not radiate. In that respect Larmor's theory has to be modified, in the light of later knowledge.

It has also to be supplemented. For early this century a step of revolutionary and surprising importance was made, in the detection by Prof. Planck, of Berlin, of the quantum. We all hear about the quantum: we do not hear so much about the quantum as we do of Einstein, but it is, nevertheless, a very important thing in physics. It is that curious and unexpected phenomenon which prevents free and easy radiation of energy, and causes it to be emitted in packets or bundles in a numerable manner—by which I mean that the packets can be counted—and that no fraction of a packet can be got. It is rather like buying postage stamps. You may buy any number but not half a stamp ; or it is like coins of the lowest denomination, you can pay with any number of them but you cannot have fractions; so that it seemed as if energy was given to us in quantum units or packets, but no fractions—if you could not have a whole quantum you could not have anything. If an atom cannot emit a whole quantum of energy it does not lose any energy at all. But quanta are not of fixed size: their magnitude depends on radiation frequency. A long-wave quantum is quite small, and it is easy to emit such ; but to emit a short-wave or really big quantum demands violence or high temperature. The quantum thus brilliantly discovered in physics is a fundamental constant of Nature, and we now see that it must depend somehow on the properties of the atom. Although a great deal has been said in the other direction I contend there is no discontinuity in the ether or in energy or in time, but there is discontinuity in a group of electrons, and in the atoms which are built of electrons, and, accordingly, in atomic theory the quantum is of fundamental importance, and in the hands of Prof. Böhr, of Copenhagen, has begun to solve the secret of atomic constitution. Prof. Böhr's work is well known to those who know it, and too elaborate for anyone else. But it is the needed supplement to Larmor's theory, and the quantum may be said to be the first vital and fundamental discovery in the region of electromagnetism since Clerk Maxwell ; for all the rest of the theory of electricity, magnificent and extensive as it is, may be regarded as a legitimate development of those astonishing equations which contain within themselves a great part of the secret of the ether, except in so far as it is modified and sophisticated by the presence of the discontinuities of matter.

CHAPTER X

Earth Transmission

IN the early days of Hertz waves—1888 to 1894—when we were engaged in demonstrating the reality of electrical oscillations in the ether, that is, in free space, we avoided earth transmission as being unfair and deceptive, suggesting the conclusion that the disturbance was being transmitted by the conducting power of the earth, rather than by the insulating power of the ether.

Thus, for instance, if a Hertz vibrator or a discharging Leyden jar was attached to the gas-pipes of a building, it was easy enough to get disturbances in neighbouring buildings, and to light the gas at a distant jet by turning it on and bringing a finger or piece of metal near it. It was possible also, and more interesting, to get sparks in unexpected places, as, for instance, between gas and water-taps which happened to be near each other in a basement, when the discharge circuit of a condenser was completely insulated in a room above—which is indeed a modification of the experiment that Joseph Henry conducted at

Washington D.C. long ago, though at that date he could not know the meaning of it.

So long, therefore, as we were experimenting with wireless waves mainly for theoretical and, so to speak, optical purposes, verifying Clerk Maxwell's theory of light — reflecting, refracting, and polarizing the waves by suitable appliances, and seeing how far and by what means they could enter closed spaces—it was natural and proper to have the main oscillator insulated from the earth, and avoid anything that could be suspected in the direction of earth conduction.

But when Senatore Marconi, in 1896 and onwards, applied these waves to practical telegraphy, his object was to get the signals at a distance, no matter by what means they went : and he therefore naturally and properly employed earth conduction for all it was worth, making a good earth-connexion at both sending and receiving stations, so that the earth became part of the oscillator. Thus began the Marconi earthed aerial, which, as we all know, is very effective for big distances.

We were all fully aware that the effect of this would be to dichotomize the waves, which had been depicted diagrammatically by Heinrich Hertz, and to assist them to run along the surface of the ground so far as it was

conducting. It was natural therefore to expect
and to get better results over sea than over
land.

So also if we wanted to get signals from a
coherer inside a closed metallic chamber, we
found we could do it by allowing some metallic
conductor to enter that chamber, provided it
was insulated from the chamber at its place
of entry. The entry of a gas-pipe, for instance,
into a metal-coated room, had the virtual
effect of turning the room inside out, and of
thus eliminating its screening property ; though
without such introduced conductor a metallic
enclosure was a complete stopper of ether waves.
But we found that the merest chink, or even
a bad join in the metal-coating of a chamber,
allowed some of the waves to penetrate ;
though a round hole was not equally effective,
unless a bit of gutta-percha covered wire was
put through the hole, so as to act as a sort of
speaking-tube or conveyer of the waves into
the interior.

We also suspected that any stray conductors,
like wire fences or buried lodes or any other
conducting material, would help to transmit
the waves. And Dr. Alexander Muirhead
applied to a cable company for permission to
make connexion with the outer iron sheathing
of a cable, in order to see whether the signals

could not thus be transmitted with greater ease. The dislike of any responsible cable company to the idea of high voltages applied to any part of their cable, prevented this experiment ; and good earth connexion was found sufficient without it.

That metallic conductors conveyed the waves was, however, well known ; for the first evidence of the existence of such waves along wires was obtained by reflecting them so as to get nodes and loops, early in 1888 and even before the publication of Hertz's great discovery of their ready transmission by free space.

How far earth-conduction assisted, or would be likely to assist, the transmission of waves between a completely insulated transmitter to a completely insulated receiver, became an open question ; and at Messrs. Muirhead's works many experiments were made of this kind. What we found was that the avoidance of earth connexion assisted the definiteness and purity of the waves, prolonging the oscillations and rendering very accurate tuning possible. We found, indeed, that earth connexion spoilt the tuning by damping the waves ; and we therefore preferred to use an aerial consisting of two insulated capacity areas, one elevated as high as practicable, and the other suspended at a fair height above the earth. We found, indeed, that the best position for the lower capacity area was at such a height that the capacity of the whole was a minimum. If the lower area were put higher, it was brought too near the upper area. If it was put lower, it was brought too near the earth. We found a position at which the tuning was sharpest, and a record of these experiments was published by the Royal Society at a later date when they were fairly complete. (See *Proceedings of the Royal Society*, vol. 82, p. 227, 1908–9.)

It was found, however, that for practical purposes the use of the earth as the lower area was simpler in practice, and in some positions was inevitable ; as, for instance, on board ship, where the sheathing of the ship made a perfect earth, and no lower area could be tolerated. It was found also that the lower area or balancing capacity was always rather a nuisance and an expense, and that even if it was constructed it was inconvenient unless buried. It was found also that the extra damping, though in itself to a certain extent objectionable, was not so deleterious as might have been expected, when a coil of considerable self-induction was employed ; since the inductance of the circuit, by prolonging the waves, could partly overpower the damping effect of the earth.

The thoroughness of earth-connexion, however, might vary in different localities. Some soils constitute a very bad conductor, others a good one. Wherever sea-water is available there is no question but that it is desirable to use it. For whether earth conduction assisted every kind of wave, even from insulated aerials, there was no doubt that it would assist when the earth was made part of the oscillating system. The only objection to it was that it was indefinite, and might have a high damping resistance. Even now it would be well to use a lower capacity area for any experiments involving really precise tuning. But for practical purposes all that was necessary was to get sufficient tuning, and to reach the greatest distance possible.

The whole subject is summarized in Prof. Fleming's treatise on " The Principles of Electrical Wave Telegraphy and Telephony " (Chap. VIII in the second edition, probably in the first edition also). And the experiments of Zenneck on different kinds of earth—that is, on the effective different kinds of soil—is there quoted and elaborated, with a citation also of further experiments by Brylinski (p. 743). The usual diagram of the dichotomized Hertzian waves is given on p. 408 in Chap. V, Section 2, where the earth is considered as a perfect conductor acting as a mirror to the upper capacity area, and as if the lower area were an equal distance below the surface. A similar discussion will be found in Pierce's " Principles of Wireless Telegraphy," Chap. XV.

How far such a condition really obtains in practice must depend on the kind of earth available ; and the distortion which the waves inevitably undergo by travelling over poor conducting soil must be learned from the experiments of Zenneck and others above referred to. (See, for instance, p. 736 in Section 14 of Fleming's treatise, Chap. VIII.)

But I see in the *Electrician* for August 11th, 1924, p. 148, that Prof. Elihu Thomson claims that this earth transmission is the really effective way in which the waves are conveyed to great distances, and are by this means enabled to go round the curvature of the earth, and reach even the Antipodes. So he concludes that any upper conducting layer in the higher parts of the atmosphere is unnecessary to explain the transmission, and that the existence of such a reflecting layer has become a superstition.

It is difficult to decide this by experiment, for we cannot get away from the upper regions of the atmosphere and determine how effective the sea alone would be without what has been called " the Heaviside layer " in the atmosphere. But it is plain that there must be a best conducting layer in the air ; since the density of the air varies, as you ascend, from its ordinary value down to absolute vacuum, and it is known that during the exhaustion of vacuum tubes, when the pressure is a few millimetres of mercury, the residual air does conduct to a degree almost comparable with the conducting power of water. It is further plain that if ether waves are confined between two strata, both fairly conducting, one above and one below, they will be kept from escaping in all directions, and will spread out in only two dimensions, thus surely economizing their power.

It may be argued that the conducting layer in the upper air is too gradual to give sharp reflection ; that is to say, that the thickness of this layer is greater than any probable wavelength, and hence that they might succeed in penetrating it. I doubt whether this is the contention. I believe that the contention of Prof. Elihu Thomson is rather that the layer is likely to be too irregular, too corrugated and uneven, to act as anything like a reasonably good mirror, even for big waves.

I should suppose that in the daytime, when the air is subject to all manner of vertical currents from the heat of the sun, such corrugations might very well occur ; but that during the night, even if there were a wind below, the upper regions of the air, with their high kinetic viscosity, might be trusted to preserve a fairly even surface. And anyhow the demonstration, by Signor Marconi's large scale experiments, of the conspicuous influence of sunlight in spoiling transmission is not readily explicable unless the atmosphere has something to do with it. If transmission depends only on earth conduction—as Prof. Elihu Thomson seems to think—one would expect signals to be as good in the daytime as in the night.

Hence, on the whole, I think facts point to a real influence of the upper atmosphere on transmission ; and whether the part played by the atmosphere can be dissected out from the part played by the earth or sea, may have to be settled by the mathematicians. And from different points of view the opinions of such authorities as Prof. Eccles, Prof. McDonald, Prof. Watson, and Dr. Chree, as well as Mr. Heaviside himself, may have to be ascertained before the question can be considered settled. I do not suppose that Prof. Elihu Thomson considers that he has settled the question, but rather that he has raised it in a more acute form, and has reopened it in all its bearings.

I have indicated sufficiently that, at the

present stage, and on merely general grounds, I have nothing dogmatic to say about it, but anyone can be impressed :—

(a) With the fact that a conducting layer in the upper air is inevitable.

(b) That such a layer, if effective, would be a great assistance in very long distance transmission.

(c) That, without it, the deleterious influence of sunlight seems rather inexplicable.

Finally, Prof. Elihu Thomson writes as if there had been undue scepticism concerning Senatore Marconi's great achievement of first getting wireless signals across the Atlantic. I know of no undue or improper scepticism about it. A hope was expressed, by myself among others, in a congratulatory letter to the Press, that further experience would confirm the result (as it conspicuously has), but caution in accepting a newspaper report of a remarkable achievement is not unprecedented, and hasty enthusiastic congratulation on the strength of such reports has not always been justified.

The transmission of ether waves round the earth is a fact of such practical and theoretical interest and importance that it has inevitably involved a good deal of discussion, and it may be well to continue to indicate some of the changes of opinion that have occurred.

H

CHAPTER XI

The Heaviside Layer

THE surprising fact that electric waves travel round the earth, instead of spreading out in straight lines like the rays of ordinary light, has set a problem to mathematicians which many have taken up and found to be of considerable difficulty. It is known that waves can be guided along conductors under certain conditions ; and, in fact, that is how ordinary telegraph signals are conveyed, whether by land wire or by cable : they travel through the insulator, but are guided by the conductor.

Conductors are opaque to waves—they cannot be penetrated ; at least, the better the conductor the more opaque it is. But a conductor can reflect waves. And if they establish a footing on its surface they can creep, or rather flash, along it with great ease, leaving a little energy behind them if the conductor is imperfect, and becoming thereby somewhat distorted, but travelling almost free from distortion if the conduction is nearly perfect.

One way, therefore, of treating the problem of long-distance transmission mathematically is to imagine the earth a perfectly conducting sphere, and find out what would happen in that case. After solving this difficult problem, the data may then be modified so as to introduce a certain amount of resistance, making the earth an imperfectly conducting sphere, as if, for instance, it were totally covered by sea-water. A third attempt, hardly one tractable mathematically, can aim at distributing land and water into continents and oceans, and seeing what happens then. That, however, is one of the empirical problems that can only be approximated to.

Another plan is to treat the subject optically, not electrically at all, and to think of waves curling round an obstacle by what is called diffraction. The laws of diffraction for small obstacles are pretty well known ; and if the earth could be treated as a small body in comparison to the size of the waves—that is, if the waves were as big as the sun or the solar system—then diffraction would be efficient, and there might be a focus or concentration of such waves at the Antipodes. But that is a quite different notion from anything appropriate to wireless telegraphy. Diffraction will not account for the curling round of ordinary ether waves. Nor is earth conduction very satisfactory.

And yet the waves do curl round, and easily

reach America, whereas if they went in straight
lines they would be going far overhead, even
for that distance. And now Senatore Marconi
appears to find that even short waves, or comparatively short waves, travel enormous distances under favourable conditions. What are
those favourable conditions ? If they were
due to earth conduction, they would not be
so likely to vary as they do. The fact that
they are capricious and dependent on sunlight
and other causes shows that the conditions
must be partly regulated by the atmosphere.

And, as is well known, Mr. Oliver Heaviside
attributed the curling-round of the waves to
the influence of a good conducting layer in the
atmosphere overhead, acting concurrently, perhaps, with the salt-water below, so that the
waves were enclosed in a stratum between two
conducting surfaces, the air effect being, on
the whole, more efficient than the earth conduction.

Everyone who has worked with vacuum
tubes with an air-pump knows that at a certain stage of exhaustion the residual air is
conducting, or at least breaks down very easily,
conveying a current and lighting up at very
small voltage ; whereas when the air is at high
pressure, or very low pressure, great voltage is
needed to drive a current through it ; but at

the best conducting vacuum, very small voltage suffices.

Now, as we ascend through the atmosphere we pass from ordinary atmospheric pressure to zero. Consequently a best conducting layer must exist. Yet a stratum of that kind is so gradual that it is unlikely to be able to serve as the layer postulated by Mr. Heaviside, even if it were sufficiently conducting. But it is well known that air can be made conducting in various ways ; notably by X-rays, and even by ultra-violet light ; also by combustion, as by flames ; and by various kinds of physical or chemical action, even by splashing water.

These agents are said to ionize the air, that is, to eject electrons from atoms ; so that electric charges are free in the air for a time, and are able to conduct, as they do in metals, where for another reason they are extremely free.

The chief ionizing factor in the atmosphere is probably the solar rays. What we get down here of sunshine has been filtered by the atmosphere. But the upper layers of the air have to stand a bombardment of the unfiltered sunlight. By ascending a very high mountain or going up in a balloon, we may experience the sunlight only partially filtered. The result is that we get first bronzed, and then blistered.

There can be little doubt that the really un-filtered sunlight would be fatal both to animal and vegetable life.

The radiation from so extremely hot a body as the sun is of a very violent character, having all the deleterious qualities of X-rays, and others in addition. So unfiltered sunlight constitutes a powerful ionizing agent.

Also it appears that the sun itself shoots off free electrons, mingled probably with positive particles. These, according to Arrhenius, would be sorted out by the earth's magnetism, the positives falling mainly at the tropics, the negatives being deflected to the poles, where they give rise to auroræ ; the opposite charges ultimately recombining, with recognized atmospheric effects and earth currents and other disturbances.

Sunlight is one of the main causes, therefore, which may give us a fairly sharply bounded conducting stratum in the atmosphere, though it may be corrugated and otherwise distorted by heat effects. And this layer it is which has been treated as the main reflector, or whispering gallery, responsible for keeping the waves travelling round the curvature of the earth, and partially preventing their escape into space.

Dr. Eccles has dealt with the theory of an ionized atmosphere very thoroughly. And on

the whole this Heaviside layer has been felt
fairly competent for its work, though admittedly
the whole subject demands extensive observation and record of experience before the theory
can be considered in any respect complete.
Like all meteorological phenomena it is complicated by a multitude of causes, and no one
single theory can adequately cover the ground.

In one of the interesting and instructive
wireless articles which Prof. Howe contributes
to the *Electrician* once a month, he comments
(in the issue of June 13, 1924, p. 720) on
what he calls " the overworked Heaviside layer "
in the upper atmosphere, and on the criticism
of it by Prof. Guinchant, of Bordeaux.

This gentleman objects that the layer is not
sufficiently conducting, for low E.M.F.s, unless
it is ionized ; and he claims that the sun cannot ionize it, for two reasons : first, because a
constant supply of electrons would soon over-
charge the earth and deplenish the sun, much
as a thoroughly insulated filament in a valve
could not continue to do its work properly ;
and, secondly, because ultra-violet light can
only ionize things when it encounters dust or
solid particles. But I suggest that Prof. Guinchant overlooks the exceedingly high frequency
of some of the radiation likely to be emitted
by a body at the temperature of the sun. Some

of it would be X-rays, competent to ionize even oxygen atoms ; and, anyhow, there is no doubt that the upper atmosphere *is* ionized : the aurora is sufficient evidence of that.

The problem of the transmission of waves round the world is a most interesting and difficult one, and certainly the last word on it has not yet been said. But few acquainted with the facts can doubt that the atmosphere is largely responsible for the possibility. It must be the main deflector for world transmission. If it is ever found that short waves are able to go round as well as long ones—and some recent statements suggest that facts are trending in that direction—then the whole question —I do not say will have to be re-opened, for it has never been closed, but the whole question will enter on a new phase.

The way in which natural conditions seem to assist long-distance wireless communication, and, as it were, unexpectedly to lend a helping hand, is rather remarkable. It is generally said that the perfect adaptation of ways and means to ends which we frequently encounter in the operations and processes of live things must be due to their long-continued adaptation through the ages and survival of the fittest.

But that explanation cannot be applicable to a recent innovation like radiotelegraphy ;

and it is interesting to find in the earth's atmosphere a favourable agent which indirectly promotes radio communication, even at enormous distances, and thus lends itself to the convenience of man, although the inception and development of the process cannot have allowed any time for adaptation and survival.

Before closing this chapter I might try to give a notion of the latest ideas on this subject of curved wave transmission. The way in which an ionized layer in the upper atmosphere would behave, has now been elaborately examined mathematically both by Dr. Eccles and Sir Joseph Larmor. That the upper air is ionized by sunlight is certain : electrons are thrown off from the atoms under the influence of light, especially light of extremely small wave-length. Such electrons may attach themselves to atoms, and by their elastic connexion confer upon the ether extra elasticity. Or, if electrons are hollow, they may cause diminished density. In either case the velocity of light would be slightly increased, both in Eccles's theory and in Larmor's,—which are really much the same, when Eccles's later papers are taken into account. Thereby the upper portions of a wave will go quicker than the lower portions ; and thus, instead of being straight, their path will be curved, somewhat as waves of sound are curved by a wind which blows them along and blows the upper portion faster than the lower. One particular curvature would enable ether waves to get all round the earth ; and Larmor has told us the conditions under which that desirable curvature can occur. (*Phil. Mag.*, December, 1924.) Hence it may now be said that the propagation of radio waves to great distances, experimentally discovered by Senatore Marconi, has been properly accounted for, and, by help of the combined influence of atmospheric properties and solar radiation, has been explained.

DETAILS THAT MAKE FOR
EFFICIENCY

CHAPTER XII

Some Points about Capacity and Inductance

THE main essentials of a wireless installation are capacity and self-induction or inductance; introduced, for a receiving station, into a collector and a detector, and, for an emitting station, into a generator and an emitter. The emitter and the collector are one and the same —the aerial. Transfer from generator to detector is usually effected by a switch. Capacity and self-induction are the essential ingredients of an aerial, and it is on them that the wavelength depends. But it is a question what value the capacity and inductance shall have, and how they shall be arranged.

It is obvious that the more open the capacity, the better will it serve as emitter or collector; hence, whatever capacity is used, it should mainly be in the aerial, for highest efficiency. Any defect of capacity in the aerial can be supplemented by a closed adjustable capacity; which, of course, is very convenient, and will always be subordinately required for tuning.

If the aerial could be arranged so as to extend to a great vertical height, its capacity would be as open as possible, and its efficiency as emitter or absorber would be correspondingly high ; for both the radiating and the absorbing power is proportional to the square of the height.

But there are practical limitations to the height convenient, so that when the greatest available height is attained, any bulk of the aerial beyond that is naturally horizontal.

In every ordinary case, however, the figure expressing the electrostatic capacity of an aerial in metres is small. It depends on the length of wire used, but is always incomparably smaller than the length of that wire. Expressed in electrostatic measure, we shall find that for an open vertical wire the capacity is about one-twentieth the length of the wire, and that the capacity of an aerial is seldom more than one-fifteenth or possibly one-twelfth of its length.

The wave-length, however, depends on the capacity and the self-induction, being indeed six times the geometric mean of these two lengths. So, for any considerable wave-length, the length representing electric capacity being small, the length representing magnetic induction must be great.

Hence, to get any reasonable wave-length, the capacity of the aerial must be supplemented, or reinforced and made effective, by a considerable amount of self-induction. But whereas the capacity area may with advantage be as extensive as possible, there is no advantage in extending or spreading out the inductance ; on the contrary, there is an advantage in compressing it into small compass, so that quite a minute coil will serve for a great wave-length.

Why should there be this advantage in constricting the self-induction coil ? Because any capacity which it possesses is useless, and, to some extent, deleterious. There is no gain in mixing up capacity and inductance. They should be kept distinct and separate. The upper part of the aerial, combined with the earth below it, should have all the capacity ; and the self-induction coil should have as little as possible. Then the wave-length has a chance of being clear and definite.

Whatever capacity exists between the turns of the coil has the effect of shunting some of the oscillation, and making it useless. The shunted portions would have any number of indefinite frequencies, and would not contribute to the main wave-length.

This has become known to practical men, and, as a result, what is called *basket*-winding has often been adopted, in order that the turns of wire may have some intervening space between them, and so not lie too close together. Of course, this has some effect in diminishing self-induction as well as capacity, since the magnetic influence of the turns of wire on each other is diminished. But the reduction in capacity is found to more than compensate this disadvantage, and it is easy enough to get sufficient self-induction by making the coil bigger.

Only, of course, then more wire has to be used for the coil; and the more wire it contains the more capacity it has. So it is evidently a question of compromise, and the best result has to be found by practice. Some capacity between the turns is inevitable; and, apart from basket-winding, we may consider how best to secure a minimum of it.

First of all, then, thin wire is indicated. From the capacity point of view, the thinner the better. The only disadvantage of thin wire is that its resistance is high. But resistance only affects the damping of the vibrations; and the vibrations are usually sufficiently persistent to cause damping to have no great importance, unless it be excessive. Damping by radiation of energy is inevitable, and moreover useful. Other damping is of no use, but it is usually small in comparison. Of course, the wire must be of the highest conductivity. But, given that, there is a gain in keeping its thickness very small, say even No. 40 s.w.g.

If in any case so much wire has to be used that its resistance does become excessive, then instead of making the wire thicker it would be better to have several wires in parallel, the said wires being very thinly insulated from each other, and then stranded or laid together.

A strand of this kind forms a very perfect conductor for high-frequency oscillations, inasmuch as every part of a thin wire helps to carry the current ; whereas only the outside of a thick wire is effectively conductive for an extremely high frequency of oscillation, so that the effective resistance of a thick wire is considerably greater than it would appear to be when measured in the ordinary way with steady currents and a Wheatstone bridge. Such considerations do not apply to a strand of fine wires, however thinly insulated from each other they are.

It may be said, why insulate the parallel wires from each other at all ? But it is clear that if they are in metallic communication all along their length, they virtually constitute a thick wire. The ether waves cannot then gain access to more than the combined periphery. The inner wires will be screened by the outer ones, just as the interior of a thick conductor

I

is screened. Whereas if there is any insulating material between them, however thin, the ether waves can, as it were, soak in and utilize the conducting power of all the wires. (It must be remembered that it is the ether, and not the copper, which really transmits the energy ; the function of the insulating material is vital.)

Given then as thin a conductor as suffices for the quantity of electricity to be conveyed, the expression for the capacity of such a wire shows that in order to keep it small the turns of wire in a coil had better not lie close together. They can be separated by an air space, or they might be separated by a thick cotton covering outside the real insulation—a covering as airy and uncompact as it can conveniently be made.

However that may be, and however the distance between the wires is secured, it can be allowed for in the calculation ; and the best method of obtaining the separation can be left to instrument makers.

The main consideration is to use as little wire as possible in the self-induction part of an aerial ; or, in other words, to wind the coil so as to get the maximum self-induction out of a given length of wire. This will have a double advantage. It will keep down the resistance, and it will keep down the capacity—both of which must obviously depend on the length of wire used.

So far as I know, insufficient attention has hitherto been paid to this important consideration, and I doubt if coils are often wound so as to obtain the maximum self-induction. I regard this as important, and propose to take if fully into consideration.

CHAPTER XIII

Conditions for Maximum Inductance

THE conditions under which a coil can have maximum self-induction (or inductance) for a given length of wire seem to have been laid down by the great mathematician Gauss, in or about 1865, but in what form that can have been done then, I do not know. Anyhow, Clerk Maxwell, in his great treatise published in 1873, gives a number of complete formulæ for self-induction, and clearly specifies the condition for its maximum. He evidently paid great attention to the subject of mutual- and self-inductance, being probably stimulated thereto in connexion with his early determination of the absolute value of the ohm (or British Association Unit, as it used then to be called).

The first condition is that the winding should be as compact as possible, so as to bring every part of the wire as close as may be to every other part, so that as many as possible of the lines of force due to each may thread the others. That will be achieved by making the section of the wound space in the bobbin of the coil either round or square, not oval or oblong. So much is pretty obvious because that is the most compact shape ; but it is not at all obvious how big the diameter of the coil should be, in proportion to the size of the channel which contains the winding. That is what has to be worked out mathematically.

Although the working out may be considered complex, the result can be stated with great ease. Taking the channel for the wire as square, the outside diameter of the coil must bear to the inside diameter the ratio $\frac{4}{2}\frac{7}{7}$, which for all practical purposes is the same as $\frac{7}{4}$, or $1\frac{3}{4}$. Hence the shape of the coil which gives maximum self-induction can be expressed in these figures : the breadth and depth of the winding 3, the internal diameter 8, and the external diameter 14.

We may take that as granted, and in this shape the coils employed in wireless telegraphy ought to be wound (though they seldom are), no matter whether the turns are packed close together or not. That is the best and most efficient shape ; and by adhering to this shape —other things being equal—the deleterious capacity and resistance in the coil are reduced to a minimum.

It need not be supposed that the shape must be very precisely adhered to. It is a common property of maxima and minima that a slight fluctuation on each side makes but a small difference. That shape is the ideal to aim at, but some variation is allowable.

For instance, suppose, having got one coil, we want to put another alongside it in series with it, the self-induction will be immensely increased by an amount which is quite well known if the positions are given. But the shape will no longer be the best. Still, the difference is not very important ; and something like the best shape can be restored by having four coils instead of two, and putting them in pairs side by side, with one pair big enough to fit over the other. Numbering the four coils 1, 2, 3, 4, it will be best to connect them together in that order, so that

THE BEST SHAPE FOR COILS.

At the left is shown the best relative dimensions for a single coil ; at the right is shown the arrangement of a number of coils that can be connected in series, and that will still keep the best shape for maximum inductance. To insure the lowest distributed capacity they should be connected as they are numbered : 1, 2, 3, 4.

the extremities of the wire, at which the greatest difference of potential will occur, are as far separated from each other as may be. The connexion 1, 2, 4, 3, or 1, 3, 4, 2, is slightly less desirable.

The effect of putting one coil outside another, instead of side by side, is only that the mean radius of the whole winding is increased somewhat ; otherwise the expression for the self-induction is the same in the two cases. It is as broad as it is long, so to speak. Or, rather, whether the length exceeds the breadth, or the breadth exceeds the length, makes no difference. That is not obvious, but so it comes out from the formula, which is symmetrical as regards length and breadth of cross-section.

The advantage of a combination of coils like this is that it enables the wave-length to be easily changed ; that is to say, it enables a coil to be selected which shall give approximately the order of wave-length required, fine adjustments being done by means of supplementary adjustable capacity, or by an adjustable separate self-induction, or both. But we will not trouble about these tuning details, which are quite well known and understood.

Although I have emphasized the value of a maximum self-induction shape, such considerations must not be allowed to override practical convenience; and, instead of packing multiple coils into a square section, it is usually much more convenient to arrange them either side by side, or one outside the other!—that is to say, to arrange them so as to form either a cylinder or a disk. And, again, such an arrangement has an advantage; for, though the self-induction will be less than it might be with a given length of wire, the terminals are thereby kept far apart, and the capacity therefore is diminished too.

Hence I do not propose to consider any arrangement for multiple coils. When we are dealing with the single coil, however, there is no question but that the best shape is as stated, viz. external diameter 14, internal diameter 8. Further details about this we will consider later.

CHAPTER XIV

The Importance of Good Contact

I AM not sure that amateurs fully realize the importance of perfect metallic connexion in every part of a receiving set. When there is plenty of power, as when one is listening to a station in the neighbourhood, a poor kind of contact may suffice. But to get the benefit of refined and accurate tuning for distant stations, we ought to realize that a tuned response begins with exceedingly small E.M.F. The whole point of tuning is that response begins with infinitesimal surgings, which, if of the right frequency, will work up by resonance to a substantial magnitude; and that if the initial infinitesimal surgings cannot occur, there is nothing to work up, and there will be no response.

Whenever we are dealing with very small E.M.F.s, as, for instance, in thermo-electric currents, perfect metallic connexion is necessary. An E.M.F of a volt or two is able to break down a thin insulating film, such as an imperceptible coat of oxide, and establish connexion after the manner of a coherer, just as an E.M.F. of 100 volts can jump across a microscopic interval; while 3,000 volts can give a millimetre spark—that is, can jump across a coarse interval of anything short of a millimetre.

But when we are dealing with the hundredth or the thousandth, or even the millionth of a volt, no such facility exists. And yet the initial surgings from a very distant station must begin at even less than the millionth of a volt. The slightest imperfection of contact therefore is sufficient to check the initial response and give tuning no chance. The wonder is that a conductor responds at all to extremely minute force. The fact that it does, shows that some of its electrons must be free from the atoms and able to be directed by the slightest suggestion of a force, as they are no doubt in a vacuum bulb.

Not only in metals, but even in electrolytes, electrons seem free. Special tests have been made to see whether electrolytes accurately obey Ohm's Law; and they do. But they could not obey Ohm's Law if an infinitesimal E.M.F. did not produce a proportional infinitesimal current. The ratio of E.M.F. to current should be constant; and as far as experiment has gone, it *is* constant in metals and electrolytes, even for the smallest forces.

But directly we deal with insulators, that is not so. They do not attempt to obey Ohm's Law. They obstruct altogether until they break down. When broken down, they conduct freely. They are said to be ionized—that is, their electrons are set free or liberated internally. But there is a critical force necessary to break them down. This applies not only to recognized insulators but to any kind of a film, a film of oil or grease for instance, or a film of oxide. Such films cannot but exist on anything exposed to the air, where dust is prevalent. They must also exist on any surface touched by the hand, or breathed upon. It is impossible to avoid such films ; and if scraped off, they will speedily renew themselves. Loose contacts, therefore, must always be suspect. The scraping action of a sliding contact may usually be trusted to remove the film, and may leave the metals in complete contact. But if cohesion is interrupted by a shake, jar, or tremor, it may not so easily renew itself.

Hence amalgamated or soldered contacts are safer. Sliding contacts are very convenient, and may often be used, but uncertain joints are always liable to give trouble ; and some of the stray noises and capriciousness from which amateurs are said to suffer can frequently be traced to this source. It is really easier to avoid troubles of this kind than to detect them when they occur.

For purposes of sending, there is no trouble of that kind. The E.M.F.s are then big enough to break down obstacles. But for refined tuning, every part of the aerial and every detail of the set should be thoroughly well joined up.

And if sliding contacts are used, the binding or clamping should be firm enough to prevent accidental disarrangement. A gentle tap breaks contact in a coherer, as everyone who used to work with such things was well aware ; and it takes an electric impulse of finite magnitude to restore connexion. No such breaks should be allowed by anyone who desires perfect reception.

CHAPTER XV

Advantage of Low Resistance and Stranded Wire

IN a receiving set intended for the reception and accurate selection of distant stations, the importance of good joints should be supplemented by a recognition of the advantages of low resistance. Persistent oscillation is only killed by resistance ; and if a conductor could be used of infinitesimal resistance, extraordinary results could be attained. Some day, perhaps, something could be done in that direction by immersing the set in liquid hydrogen, or even helium ; for at those low temperatures the resistance of metals almost disappears. Conductors become perfect, and oscillations would work up to some approach to an infinite value, even with small stimulus. Such an arrangement could surely never be more than a curiosity, and even as a curiosity it is hardly feasible at present ; but I fully expect that someone will try it in the future.

Meanwhile, we have to do the best we can with ordinary high conductivity copper. Only it must be realized that when working with short waves, and therefore very high frequency, the inner part of a wire of any ordinary thickness takes no part in the conduction. The oscillations have not time to soak or sink into the metal, and only the skin or surface contributes to the conduction. In steady currents, every part of the wire conducts equally ; the wire acts as a tube acts to water or air, except that in the hydraulic case, surface friction retards the flow a little and leaves the interior of the tube the best and most efficient part. Whereas in the electrical case, conditions are just reversed—the outside of the conductor is the best part ; the inner portion is almost useless, except as contributing to mechanical strength.

When a wire is very thin, it may be thought of as all surface. It has no interior. Hence thin wires are more efficient, weight for weight, than thick ones.

The resistance of a thick wire is not much less than that of a thin one to high-frequency currents. At the same time there must be a limit. If a wire is too thin, though the whole of it is effective as a conductor, its resistance is unavoidably high, hence the current is somewhat throttled.

To get over that we use a stranded wire, and the strands must not be in metallic contact, otherwise the interior is obliterated, since it cannot be got at except through metal.

Slight insulation, giving a path between the wires, suffices ; a coat of varnish is enough, a very thin coating of silk is ample. The point is that the strands must of all them be bathed in ether, for it is through the ether that the waves can reach them.

The propulsion of a current in a wire is effected laterally not by end-thrust but by surface propulsion, as for instance when water is propelled in a trough by moving vanes dipping into it. If the trough is completely enclosed, the vanes cannot reach the water. That is like the interior of a thick wire.

It is amazing to how small a depth really rapid oscillations sink into a wire. They sink farther into copper than into iron, for an iron wire has to be magnetized by the interior currents, and this causes so much delay that high-frequency currents keep wholly to a microscopic skin on the surface, and of course the resistance of this thin skin is very high. So, if ever a choke is required to *kill* oscillations, by high resistance, an iron wire is suitable.

For receiving purposes we want the oscillations alive, and not killed. Hence a specially efficient aerial can be made of a great number of insulated stranded wires, even as thin as No. 40.

These remarks apply especially to the leading-in wires. The aerial itself acts partly as a capacity, and for capacity these considerations do not apply ; they only apply to resistance for high-frequency currents. Similarly, all high-frequency transformers and the different leads employed should, if perfection is aimed at, be made of fine stranded wire.

CHAPTER XVI

Some Disadvantages of Reaction

WHEN two coils are coupled inductively together, they react on each other, with the result that the inductance of each is diminished and the resistance of each is increased. Resistance is never wanted; it is always a nuisance, though unavoidable. That is why coils are wound so as to give as much inductance as possible for a given length of wire—that is, for a given resistance.

Inductance confers inertia on the current, like adding a mass of lead to a pendulum bob. It makes the oscillations persist, and it enables accurate tuning. Hence anything that diminishes the inductance and increases the resistance is to that extent deleterious. But there is more objection to reaction than that.

A coil and condenser circuit, if free and uncoupled, has a definite period of oscillation of its own, and is capable of precise tuning. When coupled up to another similar circuit, its oscillations are not free. It is rather like coupling two pendulums together; they are both hampered; one tries to share its frequency with the other.

The result is that you get a kind of double vibration, something like a three-legged race. Two men run much better when their legs are not coupled together. In the latter case they interfere with each other; neither has any longer an effective will of his own; and anything like *tight* coupling is manifestly a disadvantage.

In ordinary transformers all this has to be put up with. What is wanted then is a transmission of energy from the primary to the secondary coil. And to get the maximum transmission, the coupling must be tight. The two coils become in a sense one, and the connexion thus obtained is rather like direct connexion, without a coil at all. Inductive coupling is, in fact, simply a mode of effecting connexion and at the same time giving the option of increasing the voltage by what is called " transforming up "; which is attained when the secondary coil has a great many more turns than the primary. This is not a case of reaction in the technical sense.

Indeed, in the technical sense, reaction has a still more objectionable significance. A current, magnified by a valve and high-tension battery, is made to react upon some other non-magnified part of the circuit, and thus excite magnified vibrations in that, which once more increases the vibrations in the magnified part, and these react again ; and so on, backwards and forwards, until you get a howl.

Just as when an ordinary telephone and a transmitter, short-circuited together through a single cell battery, are made to talk to each other ; the slightest disturbance in the telephone then affects the transmitter through the air, this affects the telephone through the wire and battery, and once more it reacts on the transmitter through the air, and that again acts through the wire. So that in a short time— which need be only the fraction of a second— the two set up a howl or scream of some kind, the pitch depending on the tone or tones of the telephone diaphragm.

Of course, this magnified kind of reactance gives more power ; and if the coupling is fairly loose, so that you are only on the verge of a howl, the arrangement is very sensitive. But it is not a good arrangement, and does not conduce to good tuning.

What it does conduce to, if the coupling in any way reaches the aerial, is to increase the oscillations in the aerial, turning that into a transmitter, instead of only a receiver—a transmitter, moreover, which is approximately of the right frequency, for it is its own vibrations which are worked up to a greater amplitude.

Hence the result is objurgations on the part of your neighbours, who are receiving from you instead of from the distant station they want to listen to, receiving not only the right note, but other notes near by, excited by your coupling arrangements. They cannot well tune these out because they are so near the right pitch, but it spoils their tuning ; and if you press the coupling a little closer they will receive further howls.

CHAPTER XVII

Stray Capacities and Couplings

IT is pretty well recognized now that distributed capacity in a coil, though not wholly avoidable, is undesirable and disadvantageous. Capacity should be defined and localized, and not smeared about along with resistance and inductance. The turns of a coil are intended to act inductively upon one another by magnetic induction.

They do also act on each other by electrostatic induction, which is not wanted. That is what gives distributed capacity; and hence basket-winding and other devices are employed. The separation of the wires diminishes their mutual inductance, which is bad; but it also diminishes their electrostatic or capacity induction, which is good, and the result is a compromise.

But in addition to the recognized coils, there are also capacity and inductive effects between leading-in wires and the ordinary wire connexions. None of these is any good at all, and should be kept to the minimum. We don't want capacity in a leading-in wire, we only want conduction.

We don't particularly want inductance in a leading-in wire, though we cannot help it; moreover, it does no harm. But what we certainly don't want is mutual induction and capacity between leading-in wires. And these can both be avoided to a great extent. They are no good, and though they do not do very much harm, they are better away.

To avoid them, the wires from different parts of the circuit should not run close together and parallel to each other. If they have to cross, they might cross at right-angles, being well insulated where they cross. Wires which lead to the legs or pins of a transformer or a valve should not be bunched together, even though perfectly insulated from each other. They should be separated. If they can spread out from each other for a little distance, so as not to be even parallel, so much the better. If they are separated by a few inches, their parallel running will not matter. And some careful people attach internal radiating projections to the pinholders of their valves, etc., so that the wires which lead away from the ends of their projections shall not be close together.

These are to some extent counsels of perfection; but wireless receivers are so nearly reaching perfection, and tuning is becoming so remarkably accurate, that even these trifles are worth attention. Wires should in fact be not only insulated but isolated. The nearer and more parallel they are to each other, the more they are liable to introduce undesired disturbances and spurious effects. Even insulation is not always attended to as much as it ought to be.

The minor points to be borne in mind in a good wireless set are, then :

Low resistance and perfect joints.

Stranded wire in the high-frequency portions.

Avoidance of stray capacities and mutual inductance.

Keeping away earth-connected surfaces from the immediate neighbourhood of parts in which capacity is not wanted, such as leading-in wires.

Removal of metallic masses, especially copper plates, from the neighbourhood of coils.

And most especially, good joints everywhere, no leakages or bad insulation, and highest conductivity wire.

Referring back to electrostatic induction, masses of metal near coils are apt to prove very troublesome. When a disk of metal is brought near a coil in which are oscillating currents, the disk of metal is, of course, a closed circuit, and currents—sometimes called "Foucault currents"—are induced in it. It acts, in fact, like a single-turn coil of very low resistance. And the currents may therefore be fairly strong, so that if the primary coil were conveying strong currents, the disk of metal would get quite hot.

With the kind of currents employed in wireless receivers, there would be no perceptible elevation of temperature unless extremely delicate thermometers were used. But there would be reaction. The disk of metal would be like a secondary coil, and would react on the primary.

Now when coils are thus coupled together, the effect is to diminish the inductance and increase the resistance, and therefore to put the coil out of tune if it forms part of a condenser circuit with free oscillations. The approach of the disk of metal leaves the oscillating circuit no longer free. It is virtually coupled to another circuit, and the disadvantages of reaction set in.

Tuning is not only altered but spoilt, for a double note is generated. As was said in another chapter, it is like coupling two pendulums together, or like a three-legged race.

But it may be asked : Who brings disks of metal into the neighbourhood of a coil ?

The answer is : You do, if you are using an ordinary adjustable condenser without precautions. An adjustable condenser consists of metal plates, all parallel to each other. And if they are parallel also to the plane of some coil they will react upon it. Possibly they are not very near, in which case the reaction will not be prominent.

But none of such reaction is any use, and whatever there is is bad. What is the remedy ? Either to keep the coils and the condensers far enough apart, or to arrange the plane of the coil at right angles to the plane of the plates in the condenser. Or, more accurately, to put them in what is called "a conjugate position," in which the mutual induction is zero, so that currents in one do not induce currents in the other.

There are many such zero positions. A position of zero mutual induction is obtained when lines of force due to either coil do not thread the other—that is, do not pass through the condenser plates, in the particular case under consideration.

They may dip into it, but they must rise out again, passing through the plate on both journeys. They must not pass through the plate and then return outside. In other words, they must not effectively *cut* or thread the plate as if it were a secondary coil.

This is a kind of precaution that ought to be taken by makers of sets. And if they do not happen to be aware of it, they may be arranging metal conductors near coils without realizing that they are thereby introducing spurious effects, which anyhow are no good, and which, if strong enough, will do harm.

CHAPTER XVIII

The Use of Iron in Transformers

IRON, when used as the core of a transformer or any kind of induction coil, has two chief properties, magnetization and conduction. In that it differs from any of the other ordinary metals, which practically only have the property of conduction. When a varying current circulates near ordinary metal it induces short-circuited opposite currents in the substance of that metal, and these secondary currents react on the primary circuit in a way which is most simply described as increasing its effective or apparent resistance and diminishing its effective or apparent inductance. In this respect iron has the same properties as other metals, except that it is not so good a conductor as some of them, and hence secondary induced or so-called Foucault currents are not so strong in iron as they are in copper ; but otherwise they are just the same, in kind though not in degree.

Iron, however, has the additional property of being magnetizable. But so long as these Foucault currents last, they tend to screen it from the magnetizing effect of the primary current, since they are opposed in direction to that current. They therefore delay the magnetism, and at high frequency might protect it altogether, acting as a sort of screening skin, so that hardly any magnetic lines of force are generated inside the iron. This screening action would certainly take effect at what in wireless practice is known as " high frequency." But at audio-frequencies the Foucault currents would have time to subside, killed by the high resistance of the thin skin in which they circulate ; magnetic lines of force would have time to develop, and the iron core would be magnetized and demagnetized, or reversed in magnetism, in accordance with the fluctuations of the exciting current, though with a certain amount of lag.

Of course, the Foucault currents must be kept to a minimum by subdividing the iron. It would never do to use a solid core, or a core built up of cylinders one inside the other, or of disks screwed up together so as to make a cylinder, because in either a cylinder or a disk the Foucault currents would have a free path for circulation, and the interior of the iron would hardly get magnetized at all. The core must be subdivided laterally, not longitudinally That is why it is usually built up of a bundle of thin iron wires, which, though incompletely insulated from one another (because insulation would take up valuable room), may yet be varnished, or, at any rate, slightly coated over with sufficient oxide, to prevent free electrical circulation or passage of current from wire to wire. Their longitudinal continuity is necessary for the magnetic lines of force : their lateral discontinuity is necessary for the stoppage of induced currents.

It is true that some transformer cores are made of thin sheet stampings ; but the plane of these stampings is always at right-angles to the plane of the primary coil. The stampings being in the form of disks with the centre part cut away, the windings of the primary circuit are taken through the centre hollow of the disks and back round outside ; so that the disks are continuous only in a direction at right-angles to the current, and are discontinuous in the direction of the current itself.

All this is probably well understood. Certainly it is understood by instrument makers.

But iron has another property, called hysteresis. This means that its rise in magnetism and its fall in magnetism are not quite similar. It rises, as it were, by one path, and it falls by another. The rise of magnetism, when plotted, follows what is called " the magnetization curve." The fall follows a similar but not identical curve ; so that the two curves, when plotted, enclose an area—an area something like this—when the magnetization and demagnetization are fairly complete.

Normal hysteresis curve.

If the magnetization and demagnetization are only partial, the two curves will still enclose an area, but more of this shape :—

Hysteresis Loop if the magnetization is momentary and slightly varied (diminished and increased again) at the point P.

Now wherever curves of this kind enclose an area, it means that work is done during the magnetization which is not got back during the demagnetization. There is loss or waste of energy. If the up-and-down paths were identical, there would be no loss. But when they differ from each other, it is like imperfect elasticity : you don't get back from the spring all you put into it. You never get more, and you may get less. The difference or the loss at each cycle is represented by the area enclosed between the two curves. The fatter this area is, the more the hysteresis. In fact, hysteresis may be considered as the name given to this area, the loss of energy per cycle.

Some kinds of iron have much less hysteresis than others ; but there is always some, and accordingly an iron core does involve some loss. But the advantages due to its extra magnetic lines of forces are so great as to overwhelm this loss and give us a balance of advantage, if the number of cycles is not too great.

The loss in commercial transformers at a frequency of fifty or a hundred per second is by no means insignificant. It results in heat, which is always the outcome of waste energy ; and the transformer has to be artificially kept cool. At a frequency of a thousand a second, the loss is greater ; though inasmuch as the magnetization is probably feebler, the area per cycle is likely to be less. And so for audio-frequencies, such as are used in wireless, this source of loss can easily be tolerated ; and the transformer with an iron core is more efficient, much more efficient, than one with only an air core.

But when you come to a frequency of a million a second, the slightest loss per cycle is multiplied to such an extent that it cannot be tolerated. Both things, Foucault currents and hysteresis, dissipate energy; and when even a small amount is dissipated a million times a second, it naturally mounts up. Hence high-frequency transformers must not have iron cores. An air core has neither hysteresis nor Foucault currents ; there is then no dissipation of energy, except the inevitable amount due to resistance in the wire: there is no supplementary loss. The effect of iron in a high-frequency core would be to confuse everything hopelessly. The iron would not get properly magnetized; it would be screened by its Foucault currents, nevertheless it would dissipate energy and tend to wipe or smear out the primary oscillations, destroying their features and making anything like clear speech impossible. There would not only be waste of energy, but there would be distortion. The resistance of the wire would be practically increased by the complicated reaction effects of the core.

It is rather surprising that these effects are not deleterious even in the case of audio-frequencies. It must have some bad effect, though it appears not to matter in practice. At the same time the cores of all transformers should be very carefully made, and these bad effects kept to a minimum, by special selection of the quality of iron and by thoroughly subdividing it in the lateral direction.

On these considerations is based the familiar fact that high-frequency coils are made without iron; though in low-frequency coils the use of iron is permitted and on the whole found advantageous, it should always be used with circumspection; and it seems to me possible that sets and the articulation of loud-speakers might be improved by dispensing with iron in transmitters ; for instance, by using moving coils in a steady magnetic field. Permanently magnetized iron does no harm at all. All the effects spoken of are characteristic of varying magnetism under the influence of fluctuating currents.

K

CHAPTER XIX

Contrasting Methods of Aerial Excitation

I WILL introduce this subject by an analogy.
There are two types of model or toy loco-
motives on the market : one type driven by
potential energy, the other by kinetic energy.
The first is energized by twisting india-
rubber, or tightening a coiled steel spring. This
is an example of static energy, stored in the
shape of material strain. The other type is
energized by spinning a fly-wheel, much as a
top is spun by a piece of string. The energy
thus imparted is kinetic ; and by resting the
axle of the fly-wheel on a larger wheel, the
whole thing progresses slowly like a steam-
roller, till the energy is exhausted.

The above is an example of two different
types of mechanism. But a smaller difference
can exist between the modes of excitation of
a single type. Thus take a violin string, for
instance. There are two ways of making it
sound ; one by gently bowing it, or by blowing
on it, or in some other way working up the
oscillations gradually to a sufficient intensity.
That is one way. The other is by plucking it

—that is to say, by pulling it forcibly aside till it has acquired a certain amount of potential energy, and then liberating it, so as to oscillate freely until that energy is exhausted.

A string struck by a hammer, as in a piano, belongs to the kinetic type; for the energy is imparted in the form of motion; but it is imparted very suddenly, and it virtually amounts to shock-excitation.

Thus we have three different methods of exciting a string: a pre-arranged strain, or static method, illustrated by plucking; a gradual working up of the oscillations, as illustrated by some form of friction or bowing; and the shock excitation method, illustrated by striking, as in a pianoforte. A harpist, presumably, is able to utilize at pleasure any one of the three methods. But most wind instruments depend on the gradual working-up method. Whereas drums and triangles, and other such devices—the *batterie de cuisine*, as a musician has jocularly called it—are obvious examples of percussion.

In exciting an aerial for wireless telegraphy all three methods have been employed. In the early days, working on the lines of Hertz, a spark gap was introduced into the aerial, the upper area was charged positively, the lower area negatively, setting up a strain between them, until the air between the spark knobs gave way, a rush occurred, and oscillations began. That is the steady strain or pre-arranged method—by static electrification. And on this principle many Lodge-Muirhead stations were worked. It is a very powerful method, and very difficult to tune out, since the initial jerk is rather violent. For some purposes this is a defect ; for others it is an advantage. It was found to be a defect when arranged on the Great Eastern Railway line of steamers between Harwich and Antwerp ; for though very efficient, it was rather too efficient, and the Government forts in the neighbourhood found they could not always tune us out. With better tuning devices and without earth con-nexion it could be done ; but it was admittedly not easy.

The advantage of this mode of excitation is felt when tuning-out is *not* wanted ; that is, when you desire every station within range to hear, to whatever wave-length it may be tuned. This is the case with an S O S signal ; and accordingly for distress purposes this method of excitation used to be employed on board ship, and possibly is still so employed. It should be.

Then came the percussion method of ex-citation, utilizing what I used to call a " B

spark "—that is, the rush between the outer
coatings of two Leyden jars whenever a spark
takes place between their inner coatings. In
that case the aerial was not pre-charged at all,
but was charged with a rush or a blow, by the
impact of the liberated induced charges in the
outer coats of the jars or condensers employed.
This method of shock-excitation has been used
a good deal ; and the quenched spark system
is a modification of it, since the vibrator is left
to oscillate freely after receiving a blow, like
a bell.

The third method of excitation, that by
gradual working up, is now largely employed
in various forms at continuous-wave stations.
And so far as I know it was introduced at spark
stations by Marconi in his famous 7777 patent,
the aerial being excited inductively by an
oscillating discharge in a closed circuit, to which
it was coupled. Inductive connexion at the
receiving end had been patented before, viz.
in my patent of 1897, but not at the sending
end. For though shock-excitation is in that
patent-specification clearly foreshadowed, the
continuous working-up method remained for
future development, and when introduced was
regarded as a decided improvement. For the
oscillations do not now begin with any suddenness. They are gradually worked up from zero

to a maximum, just as you may bow a tuning-fork or a bell, instead of striking it ; and thus excite a purer tone, more satisfactory to deal with, and easier to tune out when not wanted.

In this inductive method of excitation there is, in one sense, a pre-arranged static charge, at least at a spark station ; but it is not a charge in the aerial itself. The potential energy is all in a closed local circuit. It is in the spark gap of this circuit that the strain is suddenly relieved, by fracture ; and the oscillations which then begin are employed to stimulate oscillations in the coupled aerial. The spark in the primary may be quenched as soon as it has achieved its function, so that the aerial may be left free to oscillate, without being hampered by anything like tight coupling to a closed circuit ; since that, as is well known, is liable to give waves of double periodicity—that is to say, a double kind of wave instead of a single one, a wave with two peaks, both of which it is difficult or impossible to tune out simultaneously.

The valve and arc methods of excitation are representative of the continuous-wave system, like an organ pipe steadily blown from a bellows; the oscillations being varied artificially by the operator, who makes them respond to the movements of his signalling key, the key being arranged sometimes so as to give variations in pitch instead of in amplitude, as in a flute or other keyed wind-instrument.

The inductive method of excitation, whereby the energy is communicated to the antenna kinetically instead of by static strain, bears some analogy to the fly-wheel kind of model referred to at the beginning of this chapter, as contrasted with the static energy of a wound-up spring. There is initial strain in the latter, or potential-energy, case ; there is none in the kinetic mode of excitation. The act of throwing a ball is an example of a kinetic method. The liberation of an arrow from a bow, or a stone from a catapult, is an illustration of stored potential energy suddenly liberated. So is the projection of a bullet by the chemically stored energy of gunpowder. But a magnetic gun, if such a thing were ever made, would be more kinetic in its action. The two necessarily shade into one another, because suddenness is a relative term : but suddenness in a precharged antenna, charged electrically until it reaches its bursting point, is a definite fact.

CHAPTER XX

Phase Difference in Different Kinds of Coupling

TWO circuits may be coupled together either magnetically, or electrostatically, or by direct conduction ; the latter being much the same as electrostatic connexion through a condenser. Hence either of these last may be called electric coupling, while the other may be called magnetic coupling.

Taken separately, the different modes of coupling produce much the same effect, transmitting the oscillations from an open circuit, like an aerial, to a closed and resonating circuit. But if the three modes of coupling are combined in one apparatus, they tend to interfere and neutralize each other's effect, as may be explained thus :—

When a sinuous current is oscillating in a primary circuit, the E.M.F induced in a secondary circuit, depending as it does not on strength but on rate of variation, will lag a quarter phase behind the inducing current, being related to it as a cosine is to a sine.

The current excited in a secondary circuit, attuned by suitable inductance to the primary, will lag another quarter phase behind the induced E.M.F., being related to it as a minus sine is to a cosine.

Consequently the current induced in an attuned circuit of negligible resistance will be exactly in opposite phase to the primary or inducing current, being related to it as a minus sine is to a sine. And the condenser in this secondary circuit, being charged and discharged by these induced currents, will be always in opposite phase to the condenser or capacity area in the primary circuit.

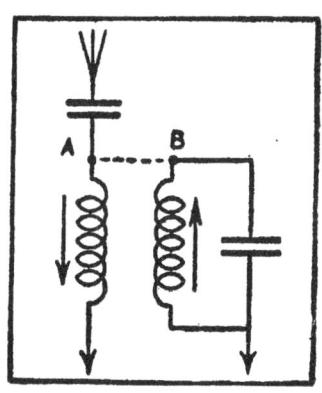

When the charge or potential of one is at maximum plus, the charge or potential of the other will be at maximum minus. That is the effect of magnetic coupling.

But the effect of electric coupling is different. In electric coupling the two con-

The arrows show primary inducing and secondary induced currents in a certain phase. The potential at A is accordingly falling, while the potential at B is rising, and *vice versa*. Consequently, if the points A and B are connected, as shown by the dotted line, the two modes of coupling will interfere and tend to neutralize each other.

densers are virtually united, so that the phase of potential or charge, in one, will correspond with the phase of potential or charge in the other. Both reach their maximum plus and maximum minus together. Hence if both couplings are employed simultaneously, they tend to interfere or neutralize each other's effect. And the couplings may be so adjusted as to produce the effect zero. The diagram on page 153 may serve to illustrate this still further.

The importance of this will be recognized when a reinforcing circuit is used ; that is to say, a free oscillator so connected as to respond only to a stimulus of one particular frequency and to exclude all others. The oscillations which it responds to are worked up to considerable amplitude by resonance provided the circuit is of low resistance and quite free from all magnetic coupling. It is stimulated by a single connector only, and its enhanced oscillations generate in its condenser a timed alternating difference of potential which can be carefully tapped off to excite grid or filament or both. Such a free circuit acts both as a selector and a magnifier, but it must not be coupled to the aerial, it must only be connected to it by a single wire or through a condenser which may be adjustable so as to vary the connexion. If doubly coupled it can act as a rejector.

CHAPTER XXI

The Grid as Traffic Regulator

SUPPOSE you have acquired a new three-valve set, consisting of :—

No. 1 valve : a high-frequency magnifier ;
No. 2 valve : a rectifier ;
No. 3 valve : a low-frequency or power magnifier.

And suppose, as sometimes happens, you find a difficulty in obtaining any but a very feeble result, or even perhaps no result at all, notwithstanding that the connexions have all been properly made, all joints and contacts good and firm, the plus and minus properly attended to, the whole high-tension battery fully connected up, and all the filaments glowing with adequate but not excessive brightness. You may then perhaps try whether you cannot get a result by turning the rheostat or regulator of the No. 1 valve to zero, so that its filament is dark and it is no longer in action : you will then be working with only two valves and without a high-frequency magnifier, and yet from neighbouring stations you may now get a result. This will not always happen, but with some sets it does. Not that this is a right way of working, but it shows what was the matter. For now, having got the other two valves to give a result, if you switch on the No. 1 valve again, you will probably find that the loud speaker again becomes faint or is silent. You ask yourself the reason of this, and conclude that the high-tension battery is too strong for that valve. Two travelling plugs are probably provided, and by moving the wander plug which feeds the No. 1 valve down a long way, to lower studs on the H.T. battery, so as to apply a much lower potential to the No. 1 anode, you may be able to get it into helpful action; and after that you can proceed as usual.

Some valves seem able to work in spite of harsh treatment, but it cannot be well to depend on that or to overstrain their capabilities. Everybody knows that the filament current must be adjusted, neither too strong nor too weak, but regulation of anode potential seems less attended to, and too little facility for this is generally provided. The result is distortion, if nothing worse. Too much reaction is no remedy, but is an additional defect. Good and pure and clearly articulated reception cannot thus be obtainable, though mere loudness can.

The fact is that the anode potential may be too high or too low for the grid potential. There is a best relation between anode and grid potential : if the anode is too high, it overpowers the grid ; if the anode is too low, it is overpowered by the grid.

Consider more closely what is happening.

Electrons are given off by the filament as negatively charged particles, and are attracted up by the positive anode. The grid stands in their way as a controller of current or regulator of traffic. The grid connected to the aerial is subject to fluctuations of potential ; it may have a steady bias, but its potential is bound to fluctuate according to the received impulses : the whole reception depends on that. When a grid is negative, it drives the electrons down or prevents their rising : when a grid is positive, it helps them upwards, and encourages them to shoot through to the anode beyond. It is the anode current which you ultimately utilize, and on which you are dependent.

But it is no use getting a strong anode current unless it is properly controlled and modified by the grid ; and the grid potential must be strong enough to perform the regulation effectively. The electrons, which are the current conveyors, must be disciplined and controlled by the grid in accordance with the received fluctuations of potential, that is, in accordance with all the voice peculiarities impressed on the ether waves by the sending microphone and the emitting valve apparatus. The anode may be so strong as to haul up all the electrons in spite of the efforts of the grid to keep them down. That is a common danger, especially with No. 1 valve. On the other hand, the grid may be so strong as not only to repel electrons when it is of negative sign, but to attract them so strongly when it is of positive sign that none or hardly any are able to escape its clutches. There are thus two opposite or alternative dangers, and the potentials must be adjusted so as to avoid them both.

It is manifest that the grids of the series of valves are inevitably of different strength : accordingly the anodes should be of different strength too. No. 1 valve receives the aerial fluctuations unmagnified, and from a distant station they may be very weak. Grid No. 2 receives magnified fluctuations, and in No. 3 they are still more magnified.

The grid potential in No. 3 or No. 4 valve may be so strong as to monopolize all the electrons to itself, not allowing a sufficient number to go through to the anode : in that case magnification will cease ; the valve will actually diminish the current which otherwise might have been obtained. Such a state of things is extremely unlikely in the No. 1 valve ; the unmagnified oscillations in the aerial are bound to be rather feeble : they are probably insufficient to excite the grid too strongly, anyhow. A very moderate potential in the anode is sufficient to do the work : indeed, a moderate potential is wisest ; for it will then not overpower the grid. And this is the basis of the plan known as " Unidyne," which refrains from applying a high potential to the anode, even to the anode of the second valve. Nevertheless, the second valve can stand a greater amount of anode potential, since the grid is already receiving a magnified impulse ; and if we are to use the third valve, a high potential to the anode becomes necessary. Without high potential you cannot expect the third valve to magnify.

ANODE POTENTIAL TOO HIGH

So far, we have mainly considered the case of a grid too strong for the anode ; it not so much regulates the traffic as stops it, absorbing too much of it into itself. But now take the converse case—the more usual case when a high-tension battery is employed—that is, when the anode potential is too strong for the grid.

The electrons given off from the filament are now rushed up violently to the anode ; and the grid placed between them in order to regulate the traffic, now stands helpless like a policeman standing in the middle of the North Western Railway trying to regulate the Scotch and Irish mails. The speeds to be dealt with are beyond the grid's control. There is plenty of current ; but the current is steady, paying no adequate attention to the fluctuations of the grid, and therefore paying no adequate attention to the received messages. Everything is in working order ; but the valve-property is out of action, the grid is no longer a regulator or controller of traffic. The remedy obviously is to weaken the anode potential. And if you want to receive from a feeble or a distant station, which is only able to make the grid voltage oscillate slightly, it will be well to reduce the anode potential a good deal. Hence doubtless it is that the Unidyne is efficient in picking up distant stations.

The anode potential of the other valves, to which grid alternations are supplied, may be higher, but still not too high. It is always important that the anode shall not overpower the grid. When listening to a strong near station that is unlikely to happen ; but when listening to a far-off station it is likely enough.

One might imagine that the feebler the impulses received, the more anode potential ought to be supplied ; whereas the fact is just the reverse ; and perhaps many amateurs overdo their high tension, especially with the early valves of the series.

To sum up :—The more distant or feeble the station listened to, the lower ought to be the H.T. potential applied to the anode of the first valve. It is possible that constructors do not allow sufficient reduction of the numbers of cells of the H.T. battery put into action on the anode, especially the anode of the first valve. Too high a potential is detrimental. The H.T. battery ought to have studs all along so as to be capable of ready adjustment down to quite a low potential—even down as low as four or six volts—and thus be made to suit different circumstances. The function of the No. 1 valve is clear reception. If it does not receive all the fluctuations clearly, subsequent magnification, so far from remedying the defect, only increases it. Given clear reception, it can be magnified by subsequent valves as much as desired. We must not depend for magnification on the receiving valve, and must not try to force it, either by reaction or by high potential or too bright a filament or in any other way.

L

A set at Egham, in Surrey, with a small aerial, that will not ordinarily receive Liverpool, is found able to do so when the potential on the first anode plate is reduced to 12 volts. With higher potential on the first plate, it can only respond to comparatively near stations, like London or Bournemouth, or, of course, to a high-power station like Chelmsford.

PART III
CALCULATIONS FOR AMATEUR CONSTRUCTORS.

CHAPTER XXII

Comparison of the Absolute Magnitudes of Capacity and Inductance

I PROPOSE to prelude some calculated and practical considerations by a little theoretical point of some interest. For wireless workers and amateurs surely like to think occasionally of the ether whose properties they are utilizing.

In electro-magnetic waves the electric energy and the magnetic energy are equal; or, in more general terms, in every wave, or system of waves, the kinetic and the potential energies are equal. This is obvious, because (at any given spot) the energy alternates from one form to the other. At one instant it is static; at the next it is kinetic. Hence the two energies must be equal.

So it is, also, with the discharge of a Leyden jar, or any other capacity area. At one instant it is charged electrically, and at the next (that is, after a quarter swing) it is momentarily discharged, and all the energy is contained in the rushing current. Then, once more, the energy piles itself up statically in the opposite direction, and then swings back again. So it is even in a swinging pendulum : the potential energy at the end of the swing is equal to the kinetic energy in the middle. So it is, also, in a vibrating spring.

Consider, then, a spring with a load on it which you can set vibrating. At the extremity of the swing the energy can be called elastic energy, or the energy of recoil. It is static. It depends on the elasticity of the spring ; it does not depend on the inertia of the load. It does not depend on inertia at all ; it would be the same if the spring were bent an equal amount and not loaded.

But now let the spring go, and consider what happens as the load is rushing past the middle position. The whole energy is now the energy of movement. It depends wholly on inertia—that is, on the massiveness of the load —it does not depend on the elasticity of the spring at all. It would be just the same for the same moving load if the spring were instantaneously abolished.

This energy may be called inertia energy, or the energy of current or movement. The elastic and the inertia energies must be equal. The spring adapts itself to them. Its rate of vibration is thereby determined. If it is a very stiff spring with a small load, it will vibrate with extreme rapidity. It must, in order that the motion energy can equal the elastic energy. If, on the other hand, it is a weak spring heavily loaded, it will vibrate very slowly; because, since the energy is small, the motion of a massive body must be slow.

All this is very elementary and simple mechanics; but now apply it to the electrical analogy. Are we to regard a Hertz vibrator or a wireless sending station as represented by a stiff spring and a light load, or a feeble spring and a heavy load? Or, again, should we not rather try to arrange it so that the spring is moderately stiff and the load moderately massive, the one being adapted to the requirements of the other, and neither being over-balanced by the other?

Now, in the electrical case, the oscillating thing is a group of electrons. They are very highly charged, but they are certainly not massive. They possess a kind of inertia due to the magnetic field which surrounds them when they are in motion. But the magnetic field due to a moving charge is but feeble, unless the charge is great and the motion exceedingly fast.

Now, the electrons, though not massive, are highly charged, and they are presumably moving very quickly. Hence their current or magnetic

energy is by no means negligible. But to bring
it up to the required amount we must magnify
it by coiling up the path of the electrons into
a close spiral, so that all the magnetic fields
reinforce each other and give a large, combined
result. In that way, by the use of a sufficient
coil, we may make the inertia what we please,
and obtain the required amount of kinetic
energy.

Now, what about the static energy? Here
we must regard the ether as strained, probably
sheared, so as to call out what is analogous to
rigidity. And the ether's rigidity is excessively
high. We know that, because of the rate at
which light travels. Its elasticity, compared
with its density, is accurately determined as
equal to the square of the velocity of light;
that is to say, the ratio of the two is excessively
great.

A very small amount of distortion will
account for a great amount of energy. But,
to make room for all the electrons which are
to take part in the discharge, an extensive area
is required. If we use only a small area, we
can hardly get any charge in it. It is like
trying to bend a very stiff spring.

A tuning-fork, for instance, can be excited
by a blow, or by a succession of timed impulses
in synchronism with its natural period, which

is practically what a violin bow does. Such a bow grips and releases a string or a spring in a synchronous, and therefore effective, manner. But a tuning-fork hardly yields to a steady pull. The amount a small force can thus bend the prong of a stiff fork is insignificant. To be able to bend it sufficiently the spring must be long, and the greater the rigidity of the material the longer it must be. That means that, to get an effective capacity area, it must be of large extent.

The aerial must be the most visible and conspicuous item in a telegraph station. On the other hand, the coil responsible for the magnetic energy may be quite small ; we might even say the smaller the better, within certain limits. The capacity area should be quite big ; we might almost say the bigger the better, again within certain limits.

There is no doubt a best relation between the size of the capacity area and the size of the inductance coil, and this relation is determined by the fact that the electric and magnetic energies must be equal. A great margin of variation is permissible, just as is the case in musical instruments, which may vary from the stiffness of a tuning-fork to the laxness of the column of air in a flute, with all manner of strings and reeds as intermediaries.

So it is with a telegraph station. One person may be working with a small capacity and a big self-induction, while another one may be working with a great capacity and a small self-induction ; and yet both may have the same period of vibration. Indeed they will have the same period if the product of capacity and inductance is the same for both. But there is sure to be a best relation between the two things which, however over-ridden in practice, it may be instructive to consider.

And it is specially instructive to realize that the great size of the aerial, as compared with the small size of the coil which is in circuit with it, is an immediate consequence of the relation which exists between the two properties of the ether, its elasticity and its density. One is incomparably bigger than the other. The ratio, in c.g.s. measure, is 10^{21}. Hence we may think that the ratio between the size of an aerial—which depends on the ether's elasticity —and the size of the little coil—which depends on the ether's density—is also of something like the order 10^{21}.

No ; it can hardly be as big as that, even with the best possible arrangement. But it is legitimate to regard that as a sort of ideal, and to emphasize the importance of a big as well as of a high aerial, and of a small, compact coil.

The size of the aerial has to be fixed by practical and often financial considerations. The size of the coil is at our disposal, and must be determined by the rapidity of vibration—that is, the wave-length that we want. And it must be adjusted so as to give this wave-length when worked in combination with the given aerial.

That, then, is the problem before us. Given an aerial of definite capacity, and required a certain wave-length, whether for receiving or for emitting—but especially we will consider receiving,—what sized coil shall we use, and what wire shall we wind it with?

CHAPTER XXIII

A Plea for Easy Specification

WHEN working with ordinary coils and condensers in the laboratory, the specification of capacity in microfarads is convenient enough, and so is the specification of inductance in terms of henries or secohms. But when working with wireless, and wave-lengths, it is convenient to have the aerial capacity, and the inductances associated with it, expressed in terms of length; because the geometric mean of those two lengths—that is, the square root of their product—gives the wave-length direct when multiplied by 2π, that is practically, for rough estimate, by 6. Six times the square root of the inductance and capacity multiplied together, is a close approximation to the wave-length: and in predetermining the inductance required for any given case this must surely be a handy rule.

It is well known that capacity in electrostatic measure is a length, and that inductance in electromagnetic measure is also a length. The truth is that in all units—that is to say, in absolute measure—capacity is really K times a length, while inductance is μ times a length. And it is natural to express the one in static measure, under the convention that K = 1, and the other in kinetic—that is magnetic— measure with the totally different convention that μ = 1. The two conventions are totally different; for the one has to do with charge, and the other with current.

The capacity of an ordinary amateur aerial is some simple fraction of its height or linear dimensions : about one-twentieth of the length of an isolated thin single wire measures its capacity. But the fraction may vary for different aerials from a twentieth to about a twelfth, as will be shown later. A microfarad is far too big a unit for convenience. A milli-microfarad, or even a micromicrofarad, has to be employed : and they are by no means con-venient. The length corresponding to a micro-farad is 9 kilometres. So a millimicrofarad is 9 metres, and a micromicrofarad, is nine-tenths of a centimetre; that is to say, 10 micromicrofarads equal 9 centimetres. So that for a rough estimate a micromicrofarad may be taken as a centimetre, though it is a trifle smaller.

On the other hand, a henry is 10,000 kilometres. So a millimicrohenry—or what is sometimes called a billihenry—is exactly

1 centimetre.   While a microhenry is 10 metres, and a millihenry 10 kilometres.

Conversion from one set of units to another is always a nuisance.   But, after all, a henry and its submultiples have no particular meaning which the imagination can seize hold of; whereas the length of a metre or a kilometre is easily imagined.   Hence it might be well to have the coils used for wireless thus marked— that is, marked in terms of length—using any unit of length that is handy for the purpose and suitable to the coil.   Thus, take an aerial of capacity 1 metre, and put a coil of 10,000 metres inductance in series with it.   The geometric mean of the two is 100 metres, and the wave-length therefore 600 metres.

The metre, as a rule, is the most convenient unit of length under the circumstances, since wave-lengths are commonly so specified.   But some people prefer to work in centimetres; and it is easy enough to remember that a billihenry is 1 centimetre.   The farad is not a convenient unit.   It was always much too big; but it can be remembered that a microfarad is equivalent to a length of 9 kilometres.   In wireless work it is certainly convenient to express capacity as a length, whether it be agreed to specify inductance also in that way, or not.

It is curious to note that a farad coupled

to a henry (or, what is more practicable, the thousandth of a farad coupled to a thousand henries) would have a slow oscillation period of six seconds, and so give a quite inappreciable wave 1,800,000 kilometres long. A microfarad connected to a henry of inductance would oscillate a thousand times in six seconds, and so generate a wave 1,800 kilometres long, which would still be feeble. Whereas a microfarad coupled to a microhenry would have a frequency a thousand times as great, and so might give a fairly strong wave of length 1,800 metres : the same wave being also generated by a millimicrofarad coupled to a millihenry ; the last being more simply expressed as a 9-metre capacity and a 10,000-metre inductance.

The intensity of radiation increases very fast as the wave is shortened. If other things were equal—which they seldom are—the radiating power from a given stock of energy would increase with the fourth power of the frequency of vibration ; so that halving the wave-length would multiply the radiating power sixteen-fold. But short waves have been usually considered less penetrating and less suitable for very long distances. Otherwise, and certainly for all near work, short wave-length or high frequency is an advantage.

CONCERNING THE SPECIFICATION OF TRANS-
FORMER AND OTHER COIL WINDINGS

It appears to be customary for instrument
makers to specify their transformer and other
windings by inscribing on them the resistance.
That is probably because the resistance is so
easily ascertained and verified. But it is not
a good mode of specification, and may lead to
misunderstanding. What we want to know
about a transformer is the number of turns
of wire in both primary and secondary, so as
to give the transformer ratio, and so as to
enable us to calculate the self-induction of each
coil, and the mutual induction between them.
These, of course, can be ascertained by experiment, even when the transformer contains iron.
But some estimate could be made of them if
the number of turns and the other dimensions
were known.   Resistance gives very little useful information.

The same is true of telephones and galvanometers. These windings too are usually specified by resistance. And there must be a
temptation to wind them with badly conducting
wire, or even some material not copper, in order
to get the high resistance more easily.   It ought
therefore to be widely known that high resist-

ance is no advantage at all. So far as it goes, it is a defect. Resistance is unavoidable in a coil wound with a great length of fine wire, but nobody wants resistance for its own sake. Resistance is only of value when *heat* is desired, as in a heating coil or a lamp filament. For all ordinary instruments the less the resistance the better. High resistance should only mean that a great number of windings have been crowded into a compact space, and the tacit assumption is that the highest conductivity wire has been used. If not, then a specification in terms of resistance is misleading. Number of turns of wire ought to be recorded on an instrument, because that cannot subsequently be ascertained. Anyone can ascertain the resistance, if they want it, without trouble, by means of a Wheatstone bridge. Either the diameter of the wire or the total length of wire used should also be recorded. Either of these quantities involves the other, if the number of turns and the mean radius of the coil is known.

Resistance is only an easy short-hand method of specification, to discriminate one coil from another, if they have all been made in the best possible way ; but without this guarantee the specification of an instrument's " resistance " may be misleading, and might lead a workman

M

to imagine that high resistance was a desideratum to be obtained in any manner he chose instead of an unavoidable condition inseparable from the other data and the properties of material.

I believe that wire as thin as No. 45 gauge can be coated with enamel as an initial insulator. If so, such wire or something rather less fragile ought to be very serviceable. And whether that wire should be wound compactly, or how far the turns should be separated from each other—either by air or by other harmless material—is a question of compromise which can be best ascertained by practical experience. If the shortest length of wire is employed, by winding it in the shape to give maximum self-induction, I doubt if it is necessary to separate the turns much ; though, of course, some insulation beyond the enamel is required. For although compact winding will give more capacity, as well as more self-induction, the reduction in the length of wire, due to the adoption of the best shape, will give a diminution of capacity—probably as much diminution as separation of the turns would give, since this would necessarily involve the employment of a greater length of wire. In practice, however, it does appear that there is an advantage in some form of basket or open-winding.

CHAPTER XXIV

On Self-Induction and its Maximum Value

THE first idea of self-induction originated with Faraday long ago, but he was quite vague about it, and called it " the electrotonic state of a conductor." It puzzled him a good deal, and he treated it almost as if it were some chemical property of the metal acquired under electrical influence. He named it " electrotonic state " in November, 1831, during his great discoveries in electro-magnetic induction generally.

The idea became rather more definite in the hands of Sir William Thomson (Lord Kelvin), who in 1853 gave the mathematical theory of electric oscillations. He perceived a sort of analogy between Faraday's electrotonic state and electrostatic capacity—only kinetic instead of static—and he therefore called it " the electrodynamic capacity of a discharger " ; in other words, he perceived that it was a constant belonging to all the wire circuit through which a Leyden jar discharged. Thus in an oscillating circuit there were the two things, both essential to oscillation : First, the electrostatic capacity of the terminal charged areas ; second, the electro-dynamic capacity of the connecting wire or discharging rod. Resistance came in sub-ordinately, as a damper or quencher of oscilla-tion, in a comparatively simple way which he thoroughly understood.

Then, later on, it was realized that just as two wires lying alongside of each other had a mutual coefficient of induction, so that the one induced currents in the other (as discovered by Faraday), each being susceptible to the rate of variation of the current in the other—so it might be said that every filament or longitu-dinal part of a single wire reacted on the other parts of the same wire ; or, in other words, that the wire was itself susceptible to the rate of variation of the current in itself. Hence it was possible to speak of not only the mutual induction of two parallel conductors, but of the self-induction of one. And so Clerk Maxwell introduced the term " self-induction," and made it quite definite and calculable. Later, Heavi-side styled it " inductance," to correspond with " resistance."

There are two ways of calculating this quantity, now commonly denoted by the letter L. One is to reckon the number of magnetic lines of force which effectively surround a wire carrying a current—the momentum, so to speak, of its magnetic field—and to call that momentum L I, where I is the strength of the current. The other is to treat the wire as if stranded, and to reckon the mutual induction of the strands on each other. This can be done by taking it as equal to the mutual induction of two parallel wires at what is called the " geometric mean distance apart "—that is to say, at a distance determined by the shape and size of the cross section of the single wire —a distance which can be reckoned as the average distance of the points in such a section from each other. It is all worked out in Clerk Maxwell's great treatise, published in 1873; and he gives an expression for this geometric mean distance for different shapes of section. It is important, because it applies not only to a single wire, but to the cross section of the wound channel in a coil. That cross section may be square or oblong or round — as when the coil is shaped like a curtain-ring.

In practice the section is usually oblong or square. It may be oblong broadways, as when one or a few layers are wound cylindrically on a tube; or oblong depthways, as when short layers are wound so as to be piled on top of each other, making a sort of disk. For a coil with one narrow dimension—that is to say, for a winding whose section is a thin oblong, whether the coil is wound horizontally or vertically—the geometric mean distance asunder of its parts is ·223, or, say, a quarter, of its larger sectional breadth. For a square section, the value is ·45 times the length of one of the sides; that is, about half the side of the square. For a circular section it is ·78 or, say, three-quarters of the radius. For an oblong section in general, the accurate expression is decidedly complicated, involving logarithms and tangents, but it may be taken as approximately a quarter of the breadth and depth of the section added together. More accurately $\frac{b + d}{\sqrt{(20)}}$, which is very nearly right. The complete formulæ will be found in Maxwell, or quoted in Prof. Fleming's comprehensive treatise, and I need not attend further to it now, because I want to concentrate on the most compact section —either a circle or a square. For it is this compactness which gives the maximum self-induction.

That, however, is not all that is necessary to be known, by any means. That only determines the shape of the channel in which the wire is wound. We must know the average size of the channel in relation to the circle of wire so formed ; that is to say, we must know

the external and internal diameters of the coil,
in terms of its sectional dimensions. Clerk
Maxwell calculates that, too, though he says
it was first worked out by the mighty mathematician Gauss, in 1867, though under what
circumstances and for what reason Gauss can
have calculated it, I do not know. It will be
instructive to some of my readers if I indicate
the manner of calculation, though those who
like may skip the algebra, which I will defer
for the immediate present. Anyhow, the *result*
is clear and definite and simple enough, and
has been given already with a diagram. The
width and depth of the channel's cross section
must be approximately three-fourteenths of
the external diameter of the coil, or three-
eighths of the internal diameter, the external
diameter being ⅝ or 1⅔ times the internal.
That determines completely the shape of the
best coil, whatever its size may be. Every coil
that we now proceed to speak of is to be of
this shape : they will differ only in size, one
will be like another magnified. But the wire
which is wound on the coils will not be magnified.
If it were, the number of turns would remain
the same, and the inductance would increase
very slowly with the additional size. It would,
in fact, in that case simply increase with the
linear dimensions, or, what is the same thing,

it would be proportional to the length of wire used.

But if the wire is maintained of constant thickness, whatever the size of bobbin on which it is wound, the inductance increases very fast as the dimensions increase. It increases not only because of the greater length of each turn of wire, but also in proportion to the square of the number of turns. If the linear dimensions are doubled, the number of turns are quadrupled, and therefore the length of wire is quadrupled; but the inductance depends on the square of the number of turns, and therefore is quadrupled twice over, making 16-fold, and the linear dimensions being doubled makes it altogether 32-fold. That is to say, increasing the size of the coil, for a given thickness of wire, increases the self-induction as the fifth power of the size. In other words, doubling all the linear dimensions multiples the inductance by 32.

The formula connecting the three things— outside diameter of coil (D), thickness of covered wire (T), and maximum self-induction (L), is as follows :

$$\frac{D^5}{T^4} = 66\cdot6L$$

Here the D, T, and L must all be expressed in the same units, no matter what those units are, and for convenience L should, therefore, in such cases, always be expressed as a length, not in such units as henries or secohms, though these are useful for other purposes.

So also it is best for wireless apparatus to express capacity as length, and not in farads or microfarads or micromicrofarads. It is much better to express it in metres, because one usually wants to employ it to calculate the wave-length. The wave-length is 2π times the geometric mean of the inductance length and the capacity length; that is, about 6 times the square root of their product. Thus suppose L is 10 kilometres and C is 1 metre, the wave-length would be 600 metres. If L is 1 kilo-metre and C is 10 metres, the wave-length is the same. If L is 16 millihenries, or 16×10^6 centimetres, and C is 100 centimetres, the wave-length will be 240,000 centimetres, or about $2\frac{1}{2}$ kilometres.

I repeat the essence of what I said before : by thus working in length units, the calculation is quite simple, and can be done in one's head, and slips of extensive magnitude can be avoided, because there is a common-sense feeling about the size of the quantities dealt with, all the time, which prevents their being accidentally taken hundreds or thousands of times too big or too small, as may easily happen when hastily dealing with meaningless units of quite unsuitable size. To measure things in farads and henries when we want the dimensions of a coil in inches, or a wave-length in metres, is not practically convenient.

The term "secohm" has gone out of use, so it may be well to explain that it was the first name given to the unit of inductance, which afterwards by international usage was styled a "henry," in honour of the pioneer work of Joseph Henry of Washington. No change was made in the magnitude of the unit: it is still 10^9 c.g.s.; but it is sometimes convenient to remember that a henry is accurately equal to a second of time multiplied by an ohm of resistance. For resistance is μ times a velocity, while inductance is μ times a length. If the inductance of a circuit ever changed at the rate of a henry per second—as a revolving armature might—it would simulate, and be equivalent to, an additional resistance of one ohm.

CHAPTER XXV

Desiderata for Inductance Coil of Receiver

IN the first place, to keep its capacity down the actual wire used should be thin, so as to expose but little surface, because distributed capacity is deleterious, so far as it can be avoided. The wire should be of the highest conductivity, but the smaller its diameter the better, so far as this desideratum is concerned. Also the shorter the length the better, since the capacity varies directly with the length. The only disadvantage of a very fine wire is that its resistance is high. But, after all, resistance does not much matter. The vibrations are damped out by radiation and other causes, and so long as the wire is able to carry the current, that will suffice. For a receiving station the current is feeble, and the thinnest wire will serve. It may be coated with silk or enamelled. And if a stranded core is employed, the enamelling of each separate strand is sufficient to keep them isolated from each other.

But it is well to wind the turns of wire not too close together. Hence a fairly thick cotton covering may be applied outside the real insulation, so as to diminish the capacity effect of each turn upon the others. The thickness of the ultimately covered wire may therefore be three or four, or even ten, times the thickness of the copper core : but I doubt if it is necessary to use a covering as thick as that. And were it not for the practical experience which has developed " basket " or open winding, I should have been disposed to advocate a close compact coil, wound so as to give maximum self-induction for a given length. In any case, maximum self-induction must be aimed at, whether the covered wire be thick or thin. I shall assume, then, that the wire to be used has an external diameter or thickness T, and that the copper core has the thickness t, and shall proceed to consider what is to be done with it. Let it be understood that with open winding T stands for the distance between the turns of wire, no matter what the covering is. It may be silk or cotton, or it may be air ; the effective thickness T is the distance of the wires apart.

Given the aerial capacity and the wavelength, or range of wave-lengths, desired, we can at once determine the self-induction, or range of self-inductions, necessary. Here is the formula, which gives the coil self-induction as the square of the wave-length divided by forty times the aerial capacity, everything being expressed in the same units of length :

$$L = \frac{\lambda^2}{4\pi^2 C}.$$

For instance, to receive a wave-length of 200 metres with an aerial whose capacity is 1 metre, which would be a likely value for a small amateur aerial, the coil to put in series with it should have an inductance comparable to

$$L = \frac{40,000}{40} = 1,000 \text{ metres };$$

that is, 1 kilometre or 10^5 centimetres or a tenth of a millihenry. It can be always remembered that π^2 is very nearly 10.

To get a wave-length of 1,000 metres with an aerial of 2 metres capacity would need an inductance

$$L = \frac{10^6}{80} = 12,500 \text{ metres };$$

that is, $12\frac{1}{2}$ kilometres or $1\frac{1}{4}$ millihenry. Twice this value would be needed if the capacity of the aerial were halved ; whereas if the wave-length to be emitted or received were doubled, using the same capacity, the inductance must be quadrupled.

Now, to get the necessary self-induction in a coil, using the smallest length of wire, we shall show in another chapter, what has already been stated, that it must be wound on a frame of the following shape and dimensions, viz. a disk coil of external diameter 14 units, of internal diameter 8 units, and with the channel for the wire a square, 3 units in the side. There remains only to determine the size of the unit which will give the required inductance for wire of given external thickness. The formula for determining the actual size of the coil's external diameter D is

$$D^5 = 66.6 \, L \, T^4$$

or

$$D = 2.31 \sqrt[5]{L \, T^4} \, ;$$

and once having determined D, the size of the coil is known in every detail, also the number of turns of the given kind of wire, and the length of wire necessary.

The use of this formula will be best illustrated by an example. Suppose the inductance required is a millihenry, that is to say, 10 kilometres or 10^6 centimetres ; and let the thickness of the covered wire (i.e. the covered wire plus the air space, if any) be 2 millimetres or $\frac{1}{5}$ centimetre ; then D^5 comes out from the above formula as $\dfrac{66.6}{625} \times 10^6$, or a trifle more than 10^5 ; and therefore the extreme diameter of the coil should be D = 10 centimetres practically, the internal diameter d will then be $\frac{8}{14}$ D = 5.7 centimetres, the breadth of the coil, or side of the square channel in which the wire is wound,

$$b = \frac{3}{14} D = 2\cdot142 \text{ centimetres,}$$

the number of turns of covered wire of this size, 5 turns to the centimetre, will be

$$n = \left(\frac{b}{T}\right)^2 = 115$$

the mean radius of a turn is $r = \frac{1}{4}(D + d) =$ nearly 4 centimetres ; and hence the total length of wire is

$$l = 2\pi n r = 27\cdot6 \text{ metres.}$$

If, to verify these approximate figures, we now use a simple formula (which is to be subsequently justified) and reckon 3 n l, we shall find that the inductance comes out 9,522 metres, which is sufficiently near the ten kilometres or 1 millihenry aimed at, for practical purposes.

Or it may be more convenient to work with inches, so far as the workshop dimensions are concerned. If we are dealing with the same self-induction we must divide 10^6 by 2·54 to bring it to inches. Or we may take as example a round number   Let the required inductance be L = 400,000 inches, while T, the thickness of the wire, $= \frac{1}{10}$ inch. Then we can reckon the external diameter of the coil, in inches, as

$$D = \sqrt[5]{\frac{66 \times 400,000}{10,000}}$$
$$= 4\cdot84 \text{ inches.}$$

Then the internal diameter will be

$$d = 2\cdot72 \text{ inches}$$

and the side of the square channel

$$b = 1\cdot03 \text{ inch.}$$

The number of turns will be $\left(\dfrac{b}{T}\right)^2$, or

$$n = 106$$

and the length of wire used

$$1 = \tfrac{1}{2}\pi n\,(D + d) = 1,260 \text{ inches or 35 yards.}$$

Checking this by $L = 3\,n\,1$ (*see* Chap. xxvii), we find it comes out right.

The result we see is not a large coil, even for so thick a covered wire. By diminishing the thickness or space between the turns of wire the coil can be much decreased in size. For if the size of the channel is given, then the use of a wire packed twice as closely, or of half the thickness, will give a 16-fold self-inductance, because it depends inversely on the fourth power of the thickness. This is indeed obvious. For if the wire is half as thick, double as many turns can be put in each layer, and there will be twice as many layers, so the number of turns altogether is quadrupled. And as the self-induction depends on the square of the number of turns, that will be magnified 16 times.

As regards size of bobbins for a given thickness of wire, we can make this statement : that doubling the linear dimensions of the bobbin for a given wire will magnify the self-induction of the resulting coil 32 times. This is not quite so obvious, but it clearly appears from the formula, since L varies with the fifth power of D, and $2^5 = 32$.

Meanwhile, the question obtrudes itself : Where does the 66·6 come from ? That is rather a long story, and must be the subject of another chapter.

N

CHAPTER XXVI

How to Calculate the Inductance of Coils

THE expression given by Maxwell for the inductance of any compactly wound coil whose diameter is considerably greater than its width, is the following :

$$L = 4\pi n^2 r \left(\text{Log.} \frac{8\,r}{R} - 2 \right)$$

Where r = the mean radius of the coil, n = the number of turns of wire, R = the geometric mean distance of the points in the section of its winding, R comes out just about one-quarter of the width and depth of the wound channel, added together (cf. p. 182). This is sufficiently accurate for ordinary purposes. Thus, suppose the width was half an inch and the depth a quarter of an inch, the value of R would be a quarter of three-quarters, or three-sixteenths of an inch.

In applying this formula it is necessary to remember that the numerator of the fraction inside the brackets, whose natural logarithm is to be taken, must be decidedly bigger than the denominator—at least ten times greater—otherwise the 2 cannot be subtracted from it without causing error or even absurdity. For log 10 = 2·3, and the balance is dangerously small even then. It is necessary, therefore, that the radius r be itself at least double or treble the greatest sectional dimension, in order that the margin may be adequate and the formula applicable.

A much more elaborate expression to cover any case is given by Maxwell, and simpler ones have now been invented and may be cited in due course.

Meanwhile, we may examine the above expression to see when it is a maximum for a given length of wire. It is plain that the length of wire wound on the coil will be the number of turns multiplied by the average length of a turn, namely $2 \pi r$; or $l = 2 \pi n r$. So the expression is :

$$L = 2 n l \left(\log \frac{8 r}{R} - 2 \right)$$

Having decided on the *shape* of the section or channel for the winding (say square), it is plain that R varies with its linear dimensions, and that therefore n varies as R^2. It also varies inversely as r for a given length of wire. Putting these things together, it follows that L is a maximum when

$$\log \frac{8 r}{R} = 3\tfrac{1}{2},$$

and that accordingly the maximum value of L
is 3 n l, which is the same as $6 \pi n^2 r$.

We will return to this extremely simple
expression for maximum inductance, L = 3 n l,
shortly. If T is the thickness of the covered
wire, and b is the side of the square in which
it is wound, it is obvious that

$$n T^2 = b^2$$

Hence the maximum self-inductance is

$$L = \frac{6 \pi r b^4}{T^4}$$

But the condition

$$\log \frac{8 r}{R} = 3\tfrac{1}{2}$$

gives us

$$\frac{8 r}{R} = e_3{}^{\tfrac{1}{2}} = \sqrt{e^7} = \sqrt{(1,096\cdot6)} = 33\cdot11$$

Or

$$\frac{r}{R} = 4\cdot14$$

Whence, for a square section,

$$r = 4\cdot14 \times \cdot45\,b$$
$$= 1\cdot863\,b$$

The outside diameter is

$$2 r + b = 4\cdot726\,b$$

The inside diameter is

$$2 r - b = 2\cdot726\,b;$$

so that

$$D = 1\cdot734\,d$$

which is just what has been frequently cited above as D : d : : 14 : 8 ; see for instance fig. 1.

Now since r = 1·863 b,
the maximum L may be written

$$6 \pi \times 1\text{·}863 \times \frac{b^6}{T^4}$$

$$= 35\text{·}1 \frac{b^6}{T^4}$$

or, using n_1 to express the number of turns per unit breadth,

$$L_m = 35\text{·}1\, n_1\,{}^4 b^6 *$$

Given the inductance required, and also the thickness of covered wire to be used, this determines b, the size of the groove which is to be filled with wire,

$$b = \tfrac{1}{2} \sqrt[5]{L\,T^4}$$

(More exactly, of course, the fifth root of 35·1 is not 2 but 2·04.)

We know that the mean radius of the bobbin is r = 1·863 b, so the external diameter of the bobbin is $\frac{14}{3}$ b, while the internal diameter is $\frac{8}{3}$ b.

Example :

Let the inductance required be 1,600 metres,

---

* Only simple arithmetic is needed to convert this into the formula given in Chap. XXV, viz. 66·6 LT⁴ = D⁵. For that is the same as 35b⁵ = LT⁴, because the fifth root of the product 35·1 × 66·6 is practically 14/3, which is the ratio of D to b.

and the thickness of covered wire half a millimetre. Then

$$L\,T^4 = \frac{1,600,000}{16} = 10^5$$

So approximately we get for breadth and
depth of section, $b = \frac{10}{2 \cdot 04} = 4.9$ millimetres ;
for external diameter, $D = 23$ millimetres ;
for internal diameter, $d = 13$ millimetres ;
for mean radius, $r = 1 \cdot 863\ b = 9.13$ millimetres ;

for number of turns, $n = \dfrac{b^2}{T^2} = 96$
for length of wire,

$$l = 2\,\pi\,n\,r = 5,520 \text{ millimetres}$$
$$= 5\tfrac{1}{2} \text{ metres}$$

If we check this by reckoning back to the
inductance as 3 n l, an expression obtained a few
paragraphs previously, we shall find it about
right. This extremely simple expression, 3 n l,
for maximum inductance, when the proper conditions are satisfied, is perhaps not of much
use for predetermination, but it is handy for
checking. It is worth considering, moreover,
whether its simplicity does not make it useful
even for the beginning of predetermination ;
and this we will do in another chapter.
     If we need to know the inductance, or the
resistance, or any other feature of a coil, ex-

tremely accurately, we must determine it by direct experiment. Its estimated value can only be a first approximation, unless excessive precautions are taken and an elaborate calculation made.

But it may be asked: " Is no notice to be taken of the thickness of the uncovered wire— the conductor itself ? " We always seem to attend to the thickness of the covered wire, or what is the same thing the distance apart of its turns, instead of to the bare copper. Well, there is a small correction to represent the concentration of the current in the core of the wire sheath. It takes the form of a small *addition* to the inductance, and Maxwell's complete expression for a ring-shaped compactly wound coil of large diameter is not merely what is quoted above, but has an additional term, so that it may be re-written thus :

$$L = 4 \pi n^2 r \left\{ \log_e \frac{8\,r}{R} - 2 + \frac{1}{n} \left(\log_e \frac{T}{t} + \cdot 12 \right) \right\}$$

where $\frac{T}{t}$ means the ratio of the thicknesses of the covered to the bare wire.

The fact that the logarithm of this ratio has to be divided by the number of turns, before using it as a correction inside the bracket, shows that it is but a small correction, even when the covering of the wire is thick or the winding open.

We will only attend to it for a coil shaped so as to give maximum inductance, and in that case the formula becomes

$$L = 3\,n\,l + 2l\left(\log_e \frac{T}{t} + \cdot 12\right)$$

Or practically

$$\frac{L}{l} = 3\,n + \log\left(\frac{T}{t}\right)^2$$

So for a coil of many turns the correction is small, but it always tends in the direction of increase.

Taking an extreme case, in which the covering is so thick that T = 10 t (or the separation of the turns ten times the diameter of the bare wire), then for a coil of 100 turns the factor with which to multiply l is not 300 but 305, when the correction is applied.

For the calculation of the inductance of coils in general, several formulæ have been elaborated by the mathematicians. Some of these are appropriate for one shape, some for another.

For compact coils, of which the dimensions of the wound channel are small compared with the diameter of the whole coil, the fundamental formula of Clerk Maxwell is probably the best. But for elongated or spread out coils—those in the shape of a long cylinder, for instance—a quite different formula is more serviceable. The Maxwellian one would then be unwieldy, and in its simple form altogether inapplicable.

For elongated or cylindrical coils an expression was worked out by Cohen.*

It applies to a single-layer coil whose length is not less than four times its diameter, and runs thus :

$$L = 4 \pi^2 n_1^2 r^2 \left(\frac{2r^2 + b^2}{\sqrt{(4 r^2 + b^2)}} - \frac{8 r}{3 \pi} \right),$$

where b is the length of the helix (i.e. the breadth of the coil), r its radius, and n_1 the number of turns per unit length.

Neglecting $(r/b)^4$, that is, taking the ratio r : b as moderately small, this can be simplified down to the following very handy formula, with $n_1 b$ replaced by n the total number of turns of wire,

$$L = \frac{4\pi^2 n^2 r^2}{b} \left(1 - \frac{8}{3\pi} \cdot \frac{r}{b} \right)$$

Or in other words,

$$L = \frac{\text{square of total length of wire}}{\text{length of coil}} \left(1 - \frac{8}{3\pi} \cdot \frac{r}{b} \right),$$

where the correction factor in brackets is of

* *See* " The Bulletin of the Bureau of Standards, U.S.A." Vol. IV, p. 385, for 1907-8. *See also* the Appendix to Professor Pierce's " Principles of Wireless Telegraphy," p. 341.

less and less importance as the helix is longer and narrower.

Example : Take a coil 1 metre long, wound with covered wire 1 millimetre thick, the diameter of the coil being 20 centimetres, so that $\frac{r}{b} = \frac{1}{10}$. The number of turns will be 1,000, and the total length of wire 20,000 π centimetres. So

$$L = \frac{(\text{length of wire})^2}{\text{length of coil}} \left(1 - \frac{8}{3\pi} \frac{r}{b} \right)$$

$$= \frac{4 \times 10^8 \, \pi^2}{100} \left(1 - \frac{8}{30\pi} \right)$$

$$= 36 \cdot 6 \times 10^6 \text{ centimetres} = 366 \text{ kilometres,}$$

which is $36\frac{1}{2}$ millihenries.

If the coil is only fairly long, and the ratio of its radius to its length be called x, then a few additional correcting terms, due to Dr. A. Russell, can be added, so that it becomes :

$$L = \frac{\text{square of total length of wire}}{\text{length of coil}}$$

$$\times \left(1 - \frac{8\,x}{3\,\pi} + \tfrac{1}{2}x^2 - \tfrac{1}{4}x^4 \right).$$

All expressions for L inevitably involve the square of the number of turns and the linear dimensions of the coil, so that the answer comes out as a length. All the rest of the formula consists of ratios of similar quantities, that is to say, of mere numbers. All such formulæ, provided there is no error, will give the right order of magnitude for the inductance, on the assumption, of course, that the core is air and not iron. But there will usually be some corrections to apply if accuracy is wanted. More accurate formulæ are given in Eccles's " Handbook to Wireless " ; but usually the accurate value is best obtained by experiment.

The value of L for a disk coil of large aperture can be reckoned from Maxwell's formula.

Or we may re-write it in terms of external and internal diameter of the disk coil thus :

$$L = \pi n^2 (D + d) \left\{ \log_e \left(16 \frac{D+d}{D-d} \right) - 2 \right\}$$

Let us apply this to a disk spiral of external diameter 6 inches and internal diameter 2 inches.

$$L = 8 \pi n^2 (\log_e 32 - 2) \text{ inches.}$$

So if there are 10 turns in the spiral,

$$L = 2,500 \times 1\cdot466 = 3,700 \text{ inches}$$
$$\text{or about 100 yards.}$$

It is remarkable how nearly the simple expression 3 n l happens to agree with this. But it won't apply at all to the case of a long cylindrical coil or solenoid. It would give a result much too big for such a coil ; though it happens to be nearly right for a disk.

A long coil may be convenient for sliding contact purposes, but it does not give maximum induction. A disk coil, on the other hand, is not usually so far away ; and indeed a properly shaped coil can be built up of disk coils.

The close addition of one disk coil to another similar one will practically quadruple the inductance, and therefore double the wave-length to be dealt with. A third coil placed close up to the other two will approximately treble the wave-length, the antenna or condenser capacity remaining the same.

One way of tuning is by varying the distance of such coils apart ; but this is commonly used only for continuous variation of reaction. Another plan is to put one coil inside another, and notch it so that their axes may have any inclination, from perpendicular to parallel, i.e. from zero to maximum mutual induction.

CHAPTER XXVII

On the Use of a Simple Formula for Maximum Self-Induction

A PERHAPS new and at any rate remarkably simple expression for the inductance of a coil wound so as to give to that inductance a maximum value for a given length of wire, has been obtained in a previous chapter, viz. that it equals the length of wire employed multiplied by three times the number of turns. Or in symbols,

$$L = 3 n l.$$

This is such a simple expression that it ought to be useful; but its applicability depends entirely on the proper conditions being satisfied. The coil must be of the right shape and size to accommodate the wire in the form of a ring of proper dimensions.

We can imagine ourselves possessed of a number of bobbins, all of the right shape but of different sizes, and may suppose that we have to choose the right size in order to give a required amount of self-induction with a covered wire of given thickness; that is to say, so many turns to the inch. Or we may consider that we have to decide on the wire suitable for winding a given bobbin in order to give the required inductance.

When I say that the bobbins are to be of the right shape I mean that they must all have 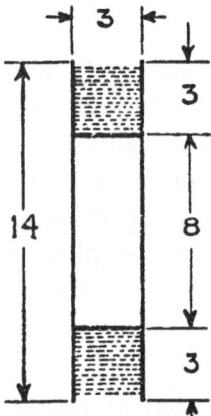 the same proportion between their dimensions and the size of the channel in which the wire is going to be wound. If the channel is 3 of any unit, say 3-eighths of an inch square, the mean diameter of the coil will be 11-eighths of an inch ; or, more completely, the external diameter will be 14-eighths of an inch, and the internal diameter 8-eighths of an inch; that is to say, the diameter of the bobbin, measured with a pair of callipers to the bottom

SHAPE OF COIL
FOR MAXIMUM
INDUCTANCE.
Size arbitrary, depending on wire used and
inductance required.

of the channel, will be just one inch.

In passing from one bobbin to another this proportion is to be maintained. Each bobbin will be just like another magnified. They will then be all of the right shape for maximum self-induction. And by suitably choosing the wire you can get any inductance you like.

The number of turns that can be wound on a given bobbin will depend on the size of the channel, which, as we know, is to be of a square section. For a given size of channel the number of turns is known. Thus suppose the wire is of such a thickness that 20 turns lie in an inch, and suppose the channel is $\frac{1}{2}$ inch wide and deep. It is obvious that we shall get 100 turns on it, ten layers of ten turns each.

The bobbin being of the right shape, if the channel is $\frac{1}{2}$ inch wide, the mean diameter of the bobbin will be $\frac{11}{3}$ of $\frac{1}{2}$ inch, that is, 2 inches less $\frac{1}{6}$. And the average length of each turn will be roughly 6 inches, more accurately $5\frac{3}{4}$ inches. So the total length of wire will be 100 times that, and the inductance 300 times that again. In other words, the product 3 n l will be 172,800 inches, or 14,400 feet.

If this inductance is near the value required, well and good. But if it is much too small, we can either choose a thinner wire for the same bobbin or select a bobbin of larger size. A small change in the thickness of the wire will make a considerable difference in the inductance. For instance, halving the thickness will increase the inductance 16-fold. A small increase in the linear dimensions of the bobbin, retaining the proportionality (as we must), will likewise make a great difference in the inductance. For doubling all the linear dimensions, without making any other change, increases the inductance 32 times. But that is hardly surprising, since one bobbin will be 8 times the weight or bulk of the other, and will be manifestly much bigger

By the use of thin wire the bobbin can be kept quite small, even for very considerable inductances. Suppose each layer of wire in a certain channel consists of 30 turns ; the total number of turns will be 900, and the length of wire that is to be used will be the required inductance length divided by 2,700 ; since that is three times the number of turns. That sort of arithmetic enables us to select a suitable wire for a given bobbin.

Another mode of writing the expression $3\,n\,l$ is $6\,\pi\,n^2\,r$, where r is the mean radius of the bobbin ; which is very nearly

$$19\,n^2\,r.$$

So if the external diameter of a bobbin were 7 centimetres, and its internal diameter 4 centimetres, so that the mean radius is $2\frac{3}{4}$ centimetres, and if 225 turns of wire are wound in its channel, being 15 to each layer of covered wire 1 millimetre thick, then (since $19 \times 2\frac{3}{4} = 52$) the inductance will be $52 \times (225)^2 = 2.6325$ million centimetres or 26·3 kilometres.

APPLICATION TO DISK COIL

The simple formula 3 n l, or its equivalent
19 n^2 r, will apply to any ring or disk coil of
fair aperture for which log 8r/R is not far
from 3·5.

For a thin disk or cylinder coil of breadth b,
the geometric mean distance of its wires from
each other is (*see* Chaps. 24 and 26)

$$R = \tfrac14 b$$

and for a disk coil

$$b = \tfrac12 (D - d)$$

while

$$r = \tfrac14 (D + d)$$

so

$$R = \tfrac18 (D - d).$$

Thus the term of which the logarithm has to
be taken in the expression for L, is

$$\frac{8r}{R} = 16 \times \frac{D + d}{D - d} ;$$

and the log of that will be

$$2·77 + \log \frac{D + d}{D - d}$$

So if $\frac{D + d}{D - d} = 2.1$, it would be just right ;

that is to say, $\log \frac{8r}{R}$ would then be just
about $3\tfrac12$.

This would mean that the outside diameter
of the disk coil would be about three times the
internal or aperture diameter ; and that is a

o

likely sort of value, modest departure from which could not affect the result greatly.

We have now incidentally justified the reckoning in Chap. 26 of the inductance of any disk coil as

$$\pi n^2 (D + d) \left(\log \frac{D + d}{D - d} + \cdot 77 \right)$$

where the ·77 represents our 2·77 (which is $\log_e 16$) with 2 subtracted from it, and if the breadth of the winding is equal to the breadth of the internal aperture, the result is 3 n l. Of course the log is the natural logarithm to base e, which is 2·3 times the log to base 10. But it must not be thought that this gives the maximum inductance for a given wire ; it would be possible to get many more turns by greater concentration.

CHAPTER XXVIII

To Estimate the Capacity of an Aerial

AERIALS can be made in innumerable shapes. But the original Marconi aerial of a single vertical, or nearly vertical, wire, suspended from a high post by an insulator, is one that is very likely to be used, with slight modifications, by an amateur, and in its simplicity it has advantages. To estimate the capacity of such an aerial, the simplest plan is to take it as one-twentieth of its length. It may be expressed in centimetres, or feet, or any units of length you please. It is unfortunately rather customary to specify it in micromicrofarads, which are not a convenient unit, though they are approximately of the order of a centimetre. Accurately, each micromicrofarad is ·9 of a centimetre—that is, 10 micromicrofarads equal 9 centimetres—which is just near enough to be confusing. Besides, centimetres are so much handier to work with. To convert a capacity expressed in micromicrofarads into centimetres, you only have to multiply it by ·9. That is to say, subtract about 10 per cent. of its numerical value—a centimetre being the larger unit of the two, and therefore a given capacity being expressed by a smaller number in centimetres.

I say, then, that the first rough estimate of a vertical wire is $\frac{1}{20}$th of its length. It will depend a little on the thickness of the wire, and still more on how near objects, such as buildings, are to it. These always tend to increase its capacity. And $\frac{1}{20}$th of its length will therefore be an under-estimate. Fleming finds that it is well to add 10 per cent. to the calculated value, in order to allow for the effect of the earth, which is inevitably not very far away from a part of the wire. This comes to the same thing as measuring it in micromicro-farads and then calling them centimetres, without reduction. In practice, it will be found that a wire suspended from any building, such as a chimney, although stretched quite free from it, but hanging down near it, will have a capacity not much less than 6 per cent. of its length, instead of 5 per cent. as above estimated for a fairly free wire.

If a wire, instead of being vertical, is horizontal, the influence of the ground is more marked ; and, at any height likely to be adopted in practice, $\frac{1}{16}$th of its length, or 6 per cent., is not a bad rough estimate.

If, however, any part of the wire is in a

building, isolated in a moderate-sized room, for
instance, $\frac{1}{12}$th of that part of its length would
be a fair guess at its capacity   Hence, if an
aerial has three portions, one part in a building, one part nearly vertical, and one part
horizontal, we might take $\frac{1}{16}$th of the length
for the horizontal, $\frac{1}{20}$th of the length for the
vertical, and $\frac{1}{12}$th of the length for the internal
portion, and add them together for the total
capacity.   Any wire coming through an earthed
tube will have a greater capacity, and to estimate that the size of the tube and of the wire
must be more accurately known.

I have said that all the above fractions will
depend to some extent on the thickness of the
wire, but they change only very slowly with
that thickness, and if the length and thickness
of the wire increased together, so that if one
was doubled the other was doubled too, no
change would be made in these fractions. They
may be taken as roughly correct for a wire
5 metres in length and $\frac{1}{10}$th of a millimetre in
diameter.   If the length is made 50 metres and
the thickness 1 millimetre, the fractions will
remain the same—that is to say, still $\frac{1}{20}$th of
the length will be a rough approximation of the
capacity for an isolated vertical wire, $\frac{1}{16}$th of the
length for an isolated horizontal wire, and $\frac{1}{12}$th
of the length for a wire isolated in a large room.

If the isolated vertical wire is much thicker, say ten times as thick, so that the 5-metre length is a millimetre thick, then instead of taking 5 per cent. of the length, we must take 6 per cent. That is the kind of difference made by a tenfold increase in the thickness.

Very often an aerial, instead of being a single wire, is a pair of wires in parallel, kept apart by distance pieces, say, a foot or two or a yard apart. In that case we might expect the capacity to be doubled. It is not quite doubled ; it is about one and three-quarters what we should estimate for each wire separately. Prof. Fleming has made experiments on the actual capacity of multiple wires, and his treatise must be referred to if more exact details are wanted. But really the capacity of an aerial ought to be measured by experiment, since that would take all the circumstances into account. It is impossible to calculate them all, and not worth while. But it is useful to be able to make a rough estimate of what the capacity will be.

Another common form is four wires arranged at the corners of a square, being kept apart by a cross-piece of wood or other material. Assuming that the wires are two or three feet apart, the combined capacity will be roughly between two and a half and two and three-quarters that of each wire separately ; and, by using a factor like that, some useful notion is obtained of what capacity to expect in a given case.

Before leaving the subject, however, we had better write down the formulæ which have enabled us to make the above rough estimates. But no simple formula like this can take account of all the varied circumstances associated with an aerial, for its capacity depends upon not only itself but upon everything in the neighbourhood. It is pretty easy to calculate the *self-induction* of a given wire, because that depends on its length, thickness, and shape. But its *capacity* is another matter. That depends on all those things, too, but it depends on other things in addition. Still, a rough estimate can be made, and the estimate is often quite sufficiently near to give knowledge of the wave-length to be expected when the estimated capacity is connected to a known self-induction.

CHAPTER XXIX

Calculation of Aerial Capacity

THE capacity of a body depends not only on itself but on its surroundings, especially on how near it is to other conductors. It also depends on the medium in which it is immersed. A condenser with glass or paraffin or mica or other insulating material between the plates has three or four times the capacity which it would have if the plates were separated only by air.

In estimating the capacity of an antenna, however, we have no medium to deal with except air, which is practically the same in this respect as absolute vacuum, or in other words, ether. For it is the ether of space and its properties on which electric capacity wholly depends. It is this which stores the energy in an elastic manner during charge, and yields it again during discharge.

The conventional unit of capacity is the farad, which is able to hold one coulomb, or one-tenth c.g.s. unit of quantity, when charged to the potential of one volt. Such a quantity of electricity can be supplied by a current of one

216

ampere flowing for one second. This quantity is enormous, from the electrostatic point of view ; and accordingly the cistern or capacity needed to hold it, at such low pressure as one volt, must be enormous too. Few of us have ever had to deal with one farad of capacity. We have to be satisfied with a microfarad, that is to say, the millionth part of a farad. But even that is too big for electrostatic purposes ; so we take the thousandth or even the millionth of that, and speak of millimicrofarads or even micromicrofarads.

Interpreting these capacities as lengths we may say :

$$\begin{aligned} 1 \text{ micromicrofarad} &= 0\text{·}9 \text{ centimetre} \\ 1 \text{ millimicrofarad} &= 9 \text{ metres} \\ 1 \text{ microfarad} &= 9 \text{ kilometres} \end{aligned}$$

while the length equivalent to a farad would reach far beyond the moon, twenty-four times the distance of the moon, or one-sixteenth of the journey to the sun.

The figure 9 which enters into these relations is rather a nuisance, but is inevitable. It comes in because the square of the velocity of light is involved ; and as that velocity in c.g.s. measure involves a 3, its square naturally contains a 9. For ordinary purposes it is sufficiently near if we consider a micromicrofarad as a centimetre, instead of ten per cent. less. It would be the capacity of a knob two centimetres, or rather less than an inch, in diameter.

On this principle we can estimate the capacity of an antenna by eye. Expressed as a length, the capacity of a vertical wire is just about one-twentieth of the height of the wire. Suppose the wire were 9 metres high, its capacity would be one-twentieth of 9 metres, that is, one-twentieth or ·05 of a millimicrofarad. To get a single-wire vertical antenna of a whole millimicrofarad capacity, it would have to be 180 metres, or about 600 feet high.

But we must now consider this matter a little more strictly, and show how to calculate the capacity on orthodox principles.

To get the capacity of an isolated wire of length l and diameter d we have only to calculate

$$\frac{1}{2 \log_e \frac{2\,l}{d}}$$

or what is the same thing

$$\frac{\cdot 217\, l}{\log_{10} \frac{2\,l}{d}}$$

It does not matter what units l and d are measured in, provided they are expressed in the *same* units, and the capacity will come out also in the same units; that is in units of length. To interpret that in millimicrofarads is easy enough in the light of what we have just been saying.

Thus, suppose an aerial wire is 60 feet long, and that the diameter is $\frac{1}{10}$ inch ; the above fraction is

$$C = \frac{\cdot 217 \times 60}{\log_{10} \dfrac{120 \times 12}{\frac{1}{10}}} = \frac{13\cdot 02}{\log_{10} 14400} = \frac{13\cdot 02}{4\cdot 16} = 3\cdot 13 \text{ ft.}$$

which is just about $\frac{1}{20}$th of the length of the wire. This equals about 1 metre ; so the capacity will be $\frac{1}{8}$th millimicrofarad. Or take another example, using the upper of the two formulæ above. Let the length of the vertical single antenna be 40 metres, and the diameter ·8 millimetre.

The capacity

$$C = \frac{40 \text{ metres}}{2 \log_e \dfrac{8000 \text{ centim}}{\cdot 08 \text{ centim}}} = \frac{40 \text{ metres}}{2 \log_e 10^5} = \frac{40 \text{ metres}}{23}$$

which again would practically be $\frac{1}{20}$th, or 5 per cent., of the length, for a little extra allowance must always be made for the inevitable partial neighbourhood of the earth. Interpreting this into conventional units of capacity we must divide by 9, and we find that it equals $\frac{40}{207}$, or say $\frac{1}{5}$th of a millimicrofarad.

If the wire is very thin, say ·008 centimetre,

its capacity would be rather less, but not much less ; we should then theoretically have to divide the length by 27 instead of by 23. Or conversely, if the wire were thick, say 8 millimetres, which could be a thin rod, we should have to divide the length by 18·4 to get the capacity. So in any practicable case we can estimate the capacity of a vertically arranged single wire as round about one-twentieth of its length, with a 10-per cent. addition for the neighbourhood of the earth.

The capacity of a horizontal wire is reckoned similarly, only then the denominator of the fraction contains four times the height above the ground instead of twice the length of the wire.

Suppose a horizontal wire at an elevation of 30 feet, of diameter $\frac{1}{40}$th inch, and of any length l, what would be its capacity ? The formula is

$$C = \frac{l}{2 \log_e \frac{4\,h}{d}} \quad \text{or} \quad \frac{l}{4 \cdot 6 \log_{10} \frac{4\,h}{d}}$$

All we need to reckon is the denominator. Now

$$\frac{4\,h}{d} = \frac{120 \text{ feet}}{\frac{1}{40} \text{ inch}} = 57600 \ ;$$

So

$$\log_e \frac{4\,h}{d} = \log_e 57600 = 10 \cdot 96$$

And the capacity is one twenty-second of its length. Every centimetre of the capacity length is practically a micromicrofarad.

But the wire for a horizontal antenna would probably be somewhat thicker ; if it was $\frac{1}{4}$ inch thick instead of $\frac{1}{40}$ inch, the denominator would become 16 instead of 22. A double wire antenna with a space of 2 or 3 feet between the wires would have a capacity about $1\frac{3}{4}$ times that of one of the wires. A quadruple wire would have $2\frac{1}{2}$ or $2\frac{3}{4}$ times the capacity of a single wire.

Consider a four-wire antenna with the wires a yard apart, each wire 1 millimetre diameter, and the whole arranged either vertically, or horizontally at half its length above the ground ; say 10 metres high and 20 metres long. Each wire if alone would be likely to have a capacity roughly estimated as its length divided by $4 \cdot 6 \times \log_{10} 40,000$, which is 21. So the capacity of the whole would be about $2\frac{1}{2}$ times the length of the antenna divided by 21, that is, $2 \cdot 4$ metres, or roughly about an eighth or ninth of the length of the wire grouping. If the wires were further separated, say by 2 yards, the combined capacity would be rather greater, and might be three times that of a single wire.

I might take a few examples to illustrate that the capacity of a wire antenna does not depend very much upon the thickness of the wire. Consider a vertical or isolated wire 20 metres long. If its thickness is ·04 centimetres or $\frac{2}{5}$th millimetres, its capacity reckoned by the formula will be :

$$C = \frac{1000}{\log.\dfrac{2000}{·02}} = \frac{1000}{\log 10^5} = \frac{1000}{11·513} = 86·85$$

or say 90 centimetres, which is the same as 100 micromicrofarads.

If the thickness is $\frac{1}{10}$th of the above, say $\frac{1}{25}$th millimetre, the denominator of the above fraction will be 10^6 instead of 10^5, and the capacity comes out 72 centimetres, instead of 90.

If, on the other hand, the wire is 4 millimetres thick the denominator is 10^4, and the capacity comes out 110 centimetres.

I will take one more example, with the length expressed in feet and inches, to show that it does not matter what units are employed, so long as we deal with them in common-sense fashion. Let l be 110 feet and d be ·1 inch. Then

$$\frac{2\,l}{d} \text{ will be } \frac{220 \times 12}{·1} = 264,000$$

So, taking the log of this as the denominator of the fraction whose numerator is the length, we find that the capacity comes out about 5½ feet, or, with allowance for earth neighbourhood, 6 feet. Every foot being 33 micromicrofarads, we can think of this as the equivalent of an isolated sphere 12 feet in diameter or as ·2 millimicrofarad.

GENERAL CONSIDERATIONS

Capacity can always be expressed in linear measure, but the etherial constant K has always to be understood or inserted if any rational meaning is to be attached to the length specification of capacity. But as K is unknown, it is conventionally taken as unity, when electrical quantities are specified on the electrostatic system. It is convenient, then, to express capacities as lengths, and inductances also as lengths, for then the square root of the product of these two lengths, that is, their geometric mean, gives a length proportional to the wave-length, and only requires multiplying by 2 π in order to give the wave-length itself in the same length units, whatever they are.

But it will always be found that the capacity is expressed by a small length, while the inductance is expressed by a big one. One may be in metres, while the other may be in kilometres. But whether they are expressed in metres, or centimetres, or feet, or yards, matters nothing, so long as one always writes down the unit of measurement after the figures. Units of measurement ought not to appear in algebraical equations or expressions, but should always be written after arithmetical figures, except when those represent a pure number. This would avoid a great deal of confusion and trouble, and is a matter not sufficiently attended to. Hence the emphasis here laid upon what ought to be an elementary and fundamental consideration. Unfortunately mathematicians who are not physicists do not always agree with this, and, like the late Mr. Todhunter, think that an algebraic expression is incomplete unless the units are stated. Physicists know that the units ought *only* to be stated in connexion with arithmetical specifications. The velocity v, for instance, is complete in itself, and is the same thing whatever units are used ; whereas a velocity 36 means nothing, unless feet-per-second, or miles-per-hour, or centimetres-per-second, are explicitly added. To say that the height of a post is 50, means nothing. But to take the height of a post as h is quite correct and complete. The post is the same height whether it be measured in feet or centimetres or anything else. So also the velocity of light is the same velocity, however expressed, and can be written completely as either *v* or *c*, according to taste.

Units should be expressed after figures, but not after algebraic symbols ; except when these are used in some shorthand technical formula, which is essentially an arithmetical one and must be so treated.

When writing down a capacity, always say what units are intended, otherwise the specification may be misinterpreted many hundred-fold. A microfarad is equivalent to a length of 9 kilometres.

This might be learned by heart.

P

CHAPTER XXX

On the Damping of Vibrations by Coils of Wire

ANY open oscillating circuit transmits its energy to the ether, and so radiates it away into space. And if a circuit consisted of two capacity areas separated by a long wire or rod, it would be an exceedingly powerful radiator, getting rid of practically all its energy in two or three swings ; which would therefore rapidly diminish in amplitude and be unsuitable for tuning or any precisely resonant effects. To prolong the oscillations we must introduce inertia in the form of self-induction, so as to make it like a heavy pendulum instead of a light one. It will then swing for a much longer time. It will not radiate so powerfully, because it will conserve its energy to some extent : in other words, the damping coefficient will be diminished, so that if left to itself it would continue swinging twenty or thirty, or even more, times ; hence if connected to a continuous wave maintainer it will be kept in vibration with but little power—at least if attuned to the right frequency.

So, the damping by radiation being diminished, the damping by wire-resistance comes into consideration. For when a current runs through a wire it inevitably wastes some energy in the form of heat, especially if the wire is exceedingly thin. Any straight conductors should therefore be fairly stout, so as not to damp the vibrations out too much ; and they are preferably made of stranded thin wires, so that every part of the copper can take its due share in conducting the current, the thin wires being insulated from each other by enamel or other convenient substance, and the whole being covered or insulated in any way desired, without detriment.

But the self-induction, which has to prolong the oscillations and lengthen the waves to what is wanted, must be in the form of a coil. The resistance of this coil should be a minimum, and its self-induction a maximum. A statement like that would be certain if capacity considerations did not matter. We must introduce those considerations directly. But it is obvious that if an unnecessary length of wire is used the resistance will be unnecessarily big, and the damping therefore more than need be. The question is whether thin wire will do for the coil, or whether that must be as stout as the leading-in wires.

Let us, then, consider the resistance and

inductance of a coil wound in a given channel,
or on a given bobbin, or other frame of definite
size. The damping depends on the ratio of
R to L (*i.e.* Resistance ÷ Inductance) ; and so
long as this ratio is constant the damping will
be the same. It does not depend on R alone,
nor on L alone, but on the ratio of the two.
Suppose now we fill the channel with a thick
wire, its resistance will be small, but its inductance will be small also. Whereas if we wind
it with very thin wire, we shall get on a large
number of turns ; so the resistance will be
high, but the inductance will be high also, and
we have to consider whether the ratio remains
the same. We shall find that it does, and that
whether the coil is wound with a single thick
rod, or whether it is wound with thousands of
turns of wire, the ratio is not altered. For the
resistance will depend on the square of the
number of turns, since the length of wire will
increase with n and the cross section of the
wire will diminish with n. Hence the resistance will depend on n^2. But the inductance
also depends on n^2. Hence the ratio of R to
L remains constant whatever wire is used, provided the coating is so thin that the channel
may be considered filled with copper in each
case.

With extremely thin wire the space is

largely occupied with insulation, and hence
there is a tendency for the ratio of R to L to
increase somewhat as the thickness of the wire
is diminished. But it increases very slowly,
and for practical purposes the increase is unimportant. Wherefore, although the leading
wires should be fairly substantial, or at least
not too constricted, the wire on the coil may
be reasonably thin. And any further details
about the winding must be dealt with from the
point of view of capacity, resistance being left
to take care of itself.

The way to keep the capacity in the coil
small is to wind it in a single layer, like a number
of coils wound on a cylinder. In that case we
have only the capacity of each turn upon those
on each side of it, unless the bobbin on which
it is wound is of some conducting material, in
which case the wire will form one coat of a
cylindrical condenser, and the capacity will be
far from insignificant. It is important, therefore,
to use really insulating material for the cylinder
on which wire is wound.

The better plan, though more troublesome,
is to wind the wire as a thin disk, a large number
of superposed layers of small breadth ; and by
adopting this method of winding, and support-
ing the disk without cheeks or metal frame of
any kind, we reduce the capacity to a minimum.

And we can, if we like, even separate the turns, making a sort of basket winding, or else a spiral with interspaces, such as is often used for a purely magnetic or frame collector.

To calculate the capacity effect of a coil we can treat each layer as if it were one coat of a cylindrical condenser ; and we shall find that we have not to add these capacities together. The effective capacity of the whole coil will not be more than the capacity of a single layer, because the whole difference of potential between the terminals will not be applied to any one layer, but only a fraction of it. The whole difference of potential exists between the terminals of the coil, that is, between the inner and the outer layers. So that suppose there are six layers, only $\frac{1}{6}$ of the difference of potential is applied to each, and to reckon the effective capacity we shall have therefore both to multiply and to divide by 6. Hence it is that the number of layers does not matter. All we want to know is the capacity of any one layer.

Take, then, the axial dimensions of a coil, or what may be called its breadth ; call that *b*. And take the radius of the coil, which we may call *r*. The layer forms a cylinder whose area is the circumference multiplied by the breadth ; that is, $2 \pi r b$. It only remains to reckon the distance separating one layer from the next, and this will be equal to the thickness of the covered wire minus the thickness of the uncovered wire. For an approximate estimate we can neglect the thickness of the uncovered wire, assuming that it is thin, and take the distance as the diameter of the wire, that is, twice the thickness of the covering, whether the covering be air or any other material.

Treating it in this way, we know that the capacity of a plate condenser is

$$\frac{A}{4\pi z}$$

where A is the area of either coating, and z the distance between the coatings. So in the above case this quantity will be

$$\frac{2\pi r b}{4\pi T}$$

if T is the thickness of the covered wire or the distance of the layers apart.

We will consider the order of magnitude of this capacity for a given example:

Let the breadth of a coil be 2 inches, and the mean radius of all the windings on it be 3 inches, and let the diameter of the covered wire with which it is wound be rather more than $\frac{1}{2}$ millimetre, or say $\frac{1}{40}$ of an inch. The capacity of each layer, with regard to the layers above and below, will then be

$$\frac{r\,b}{T} = 240 \text{ inches}$$

that is, 20 feet, which is comparable to the capacity of a single wire aerial 400 feet high !

A smaller coil might be wound with still thinner wire. And if the diameter of the covered wire is ·006 centimetre, even though the breadth of the channel is only 1 centi-metre, yet with the mean radius 3 centimetres, the effective capacity will be

$$\frac{3}{·006}$$

that is, 500 centimetres, or 5 metres, which is still very large—bigger than most amateur aerials. To have that capacity a single wire aerial would have to be about 100 metres long, and even a quadruple horizontal aerial with its four wires strained a yard apart would have to be 40 metres in length.

The coils here selected as examples of dele-terious distributed capacity are larger than any that would be wise to use, but they show how big this unexpected kind of capacity can be, and how we must make an effort to avoid it. We do not want the capacity of the coil to have any relation to the capacity of the aerial. The coil should be kept in its due insignificance so far as regards capacity. What we want in the coil is self-induction. Distributed capacity along the coil only introduces confusion, spoils the sharpness of the tuning, and makes precision impossible. It introduces the same kind of confusion as a submarine cable introduces into telephonic speech. The Leyden jar effect of a cable—that is, of a wire conductor separated bv an insulator from an outside coating—prevents there being any definite speed of transmission and tends to smooth out the signals and make them indefinite.

This effect in cables can be remedied by the introduction of coils at intervals, showing that coils are not in themselves deleterious. But they should always have as much inductance as possible in proportion to their resistance, so as not to introduce unnecessary damping.

The shape and size of the coils advocated in previous chapters satisfied these conditions as far as possible ; and the advantage of open or basket winding, to reduce capacity, is obvious.

The channel for the wire being 3 units square, the external diameter of the coil should be 14 units, and the internal diameter 8 units. The particular units can be adjusted to suit the desired conditions : for instance, each unit might be chosen as a couple of millimetres. Suppose the covered wire is $\frac{1}{4}$th of such unit thick ; then the number of turns in each layer of the above coil is 3 x, and there will be 3 x layers. So the number of turns is 9 x^2. The *mean* diameter of the coil is 11 units, and the average length of each turn is therefore 11 π or say 34 units ; wherefore the total length of wire will be 34 × 9 x^2. The inductance of a coil of this shape has already been shown, by general reasoning and calculation, to be 3 n l, being the maximum obtainable for that length of wire and therefore for that resistance and capacity. So the inductance comes out

$$L = 3\,n\,l = 8100\ x^4$$

of the same units of length, approximately, taking 3 × 34 as 100.

If the unit selected is, as above suggested, a couple of millimetres, and if the value of x is say 10—both these things being arbitrary— the value of L, as thus calculated, is 1·62 × 10^7 millimetres, or 16·2 kilometres, or 1·62 milli-henries.

The maximum deleterious or distributed capacitance of such a coil would be obtained, as above, by reckoning $\frac{r\,b}{T}$, which is 5·5 × 3 × x of the same unit ; so, if the unit is a couple of millimetres and n = 10, as above, the capacitance is 330 millimetres, which is a third of a metre. Not too big, and capable of reduction by basket winding.

To reckon the resistance, we should have to know the thickness of the uncovered wire, and its conductivity ; but resistance is a thing easily obtained by direct measurement, and is hardly worth pre-determination, as inductance is.

CHAPTER XXXI

The Romance of Wireless

THE romance of wireless chiefly consists in the fact that we are for the first time consciously using the ether of space as a vehicle for messages. We have long unconsciously used it—not when we speak to each other, for then we only use the air—but when we smile or wink or nod ; or, on a larger scale, when we signal by means of flags or semaphores, or when we flash Morse signals by the heliograph. We are also employing the ether (though few people know it) in ordinary telegraphy. The wire does not really convey the message, it only directs it to its destination. It does not convey the message any more than the rails convey a train. They direct it from, say, Liverpool to Euston ; but they take no part in its propulsion. If anything, they absorb energy rather than impart it ; they get, I presume, slightly heated by the process. And that is exactly what the wire does. A wire is said to transmit power ; it really absorbs power. The real transmitter is the ether round the wire.

In cable telegraphy it is the insulator which transmits, the conductor only guides. And it is not the insulating substance itself which transmits, but the fact that it is insulating enables the ether in it to transmit ; that is what we mean by " insulating." The electrons in an insulator are fixed in position ; whereas in a conductor they move freely But the power is transferred, not directly by the electrons, but by their electrostatic fields ; that is, by the space intervening between them—a space which, like every other space, is full of ether, and which in a cable is limited to the annular channel between the metallic core of the cable and its outer sheath. If the voltage applied is too high, it is the insulator that is damaged, not the core : and if the insulator is punctured, the message is stopped. Instead of going where it is intended, the power then leaks out and the current returns to its source prematurely, having achieved nothing of any use. At the distant end a little bit of the power which arrives should be utilized to give the signal in the sensitive responding instrument. The rest of it is wasted, the current returns, as we say, to earth, ultimately completing the circuit, and returning by unknown and indefinite paths to the battery or dynamo which originally propelled it on its journey.

When a rapidly alternating source is used, however, true waves can be emitted into the space surrounding the wire. Portions of energy are broken off and shot forward by their own momentum, travelling with the velocity of light ; not with the velocity of light in free space, but with the velocity appropriate to the insulating material—which may be about half the speed attainable in free space.

The existence of these waves conveyed by wires was demonstrated, and their wave-length roughly measured, in experiments made by me at the top of Brownlow Hill in Liverpool, in the University College which is now the University of Liverpool, in 1887–8 : and the theory of such waves was deduced from Clerk Maxwell's work.

Just about the same time that exceptionally great physicist, Heinrich Hertz—who died much too young—made the far more striking discovery that these waves could be generated and detected in free space, without any guiding wire at all. Electricity has momentum as well as elasticity ; and these two requisites suffice for the transmission of waves.

Electric and magnetic effects occur together ; they are inevitably associated. The electric part gives the elasticity, or the recoil ; the magnetic part gives the inertia, or the momen-

tum. And combining the two, as Clerk Maxwell showed, we have all the pre-requisites for
the propagation of waves.

It is by this means that we see : it is thus
that we receive light from the sun. In the sun
the ether is set quivering by the vigorous oscillation of its electric charges. The waves travel
out in all directions into space ; and a minute
fraction of them is utilized to stimulate the
sensitive receiving instrument located in our
eyes. Thus the human race and animals have
all along been utilizing the properties of the
ether to convey information. Information is
conveyed, not only about the source which
emitted the light, but about all the other
objects upon which the light falls, provided
some of it enters the eye. It is thus that we
are aware of the clouds and the trees and the
grass and all the beauties of a landscape. The
eye is our etherial receiving instrument : it is
unconscious wireless telegraphy, an unconscious
utilization of the ether.

We attend to the illuminated objects ; we
attend not at all, as a rule, to the vehicle by
which the information comes. We hardly know
that we get the information in that way.
Through a long course of evolution and ances-
tral habit, the operation seems so direct and
obvious that we fail to recognize it as a

process at all. We receive the ether waves shimmered off from the apples on a tree, as from so many signalling stations : we ignore the waves and the ether, we do not contemplate the process, we merely say we " see the tree."

So it is with everything we see. The sense of sight is so familiar, so constantly in use, that the mystery of it escapes us. We seem outside the region of romance, among the commonplace things of everyday life. But seen rightly, these commonplace things are full of mystery, full of romance, and contain far more problems than we have as yet completely solved. That information can thus be conveyed to the mind, with elaboration of the utmost significance of detail, is really overwhelmingly astonishing when we fully grasp the idea. But familiarity has bred a kind of contempt, and we fail to realize its wonder.

But now we have taken a further step, not in any way more remarkable, but with more understanding of what we are doing, under conditions therefore which are more liable to attract our attention and stimulate our surprise. Instead of the eye as detector, we construct artificial instruments, whereby we have to deal with much longer waves, though otherwise of precisely the same character as those with which the eye is competent to deal. We translate those signals into auditory vibrations, that is, vibrations of plates and of the air, and so make them accessible to the ear instead of to the eye. And we have constructed other ingenious artificial arrangements for producing and modifying those waves in accordance with the vibrations of the human larynx. The sending and receiving stations are different from those that have arisen in the course of evolution : they are artificially constructed, and accordingly are better understood ; but in essentials the process is the same. Waves are generated, and then modulated, in the ether ; they travel out in all directions ; and some small fraction of them is utilized to stimulate the receiver, and produce what we call a sound. We have, in fact, made the ether " talk," or, rather, transmit speech. Hitherto we only used the air, as when we shout to a distant person. That, too, involves much mystery—that a thought can be conveyed by the mere vibra-tions of the air. But now we use the ether, a much more perfect and omnipresent medium ; and thereby we are enabled to reach distances before inaccessible. Through it we can, as it were, speak round the world. It is certainly an amazing achievement : and all honour to those who have made it possible—a group of

Q

co-operative workers, whose names to the wireless amateur are household words to-day.

Humanity has attained a new instrument, a new weapon, a new mode of conveying thought, a new method of intercourse between the nations, almost annihilating space. Surely it will be a weapon of peace! Surely it will promote better understanding! Isolated groups of men develop insular prejudices, fail to understand other groups : jealousies and misunderstandings arise, until they lead to the horrors of internecine strife and wholesale slaughter. This is not civilization. This is below the level of civilized humanity. This is a diabolical perversion of the stage at which we ought to have arrived.

But there is nothing perverse in being able to talk to each other more freely. The commonwealth of nations often called the British Empire is peacefully disposed. They with our cousins across the Atlantic constitute the same race speaking the same language ; they are now welded together by this new method of communication. And if they only set their face to the right, and determine to preserve the peace of the world, they will succeed. The ether welds the worlds together into a cosmic system of law and order. Let it weld all humanity together, so that they can face their common difficulties in a spirit of co-operation and mutual trust ! " We beseech Thee to give to all nations unity, peace, and concord."

Finally, it may not be impertinent to suggest that the matter of our communications should be worthy of the manner, that the educational facilities of broadcasting should be utilized, that as we increase in intelligence we can increase in wisdom too. We can realize better our place in the scheme of existence, and not be satisfied with merely material achievements. Our achievements are now not only material, they are ethereal also. Spiritual development should keep pace with the others. So I conclude with the warning, expressed long ago by Mrs. Browning, with remarkable foresight and poetic anticipation of what then must have seemed distant dreams, but which now, by submarines and aviation and wireless, have been in some manner accomplished :—

" If we trod the deeps of ocean, if we struck the stars
    in rising,
    If we wrapped the globe intensely with one hot electric
        breath,
    'Twere but power within our tether, no new spirit-power
        comprising,
    And in life we were not greater men, nor bolder men in
        death."

INDEX

245

PRINTED BY
CASSELL & COMPANY, LIMITED,
LA BELLE SAUVAGE, LONDON, E.C.4
F 75.525